Manuals for the Bench

E. CARAFOLI, G. SEMENZA (Eds.)
Membrane Biochemistry. A Laboratory Manual on Transport and Bioenergetics (1979), 175 pp, ISBN 3-540-09844-5

A. AZZI, U. BRODBECK, P. ZAHLER (Eds.)
Membrane Proteins. A Laboratory Manual (1981), 256 pp, ISBN 3-540-10749-5

A. AZZI, U. BRODBECK, P. ZAHLER (Eds.)
Enzymes, Receptors, and Carriers of Biological Membranes. A Laboratory Manual (1984), 165 pp, ISBN 3-540-13751-3

A. AZZI, L. MASOTTI, A. VECLI (Eds.)
Membrane Proteins. Isolation and Characterization (1986). 181 pp, ISBN 3-540-17014-6

N. LATRUFFE, Y. GAUDEMER, P. VIGNAIS, A. AZZI (Eds.)
Dynamics of Membrane Proteins and Cellular Energetics (1988), 278 pp, ISBN 3-540-50047-2

U. BRODBECK, C. BORDIER (Eds.)
Post-translational Modification of Proteins by Lipids. A Laboratory Manual (1988), 148 pp, ISBN 3-540-50215-7

J. F. T. SPENCER, D. M. SPENCER, I. J. BRUCE
Yeast Genetics. A Manual of Methods (1989)
104 pp, ISBN 3-540-18805-3

Yeast Genetics
A Manual of Methods

By J. F. T. Spencer
D. M. Spencer · I. J. Bruce

With 10 Figures

Springer-Verlag
Berlin Heidelberg New York
London Paris Tokyo

Dr. JOHN F. T. SPENCER
Dr. DOROTHY M. SPENCER
Department of Life Sciences
Goldsmith's College
Rachel McMillan Building
Creek Road
London SE8 3BU
Great Britain

Dr. I. J. BRUCE
Thames Polytechnic
Department of Biology and
Environmental Health
Wellington Street
London SE18
Great Britain

ISBN 3-540-18805-3 Springer-Verlag Berlin Heidelberg New York
ISBN 0-387-18805-3 Springer-Verlag New York Berlin Heidelberg

Library of Congress Cataloging-in-Publication Data. Spencer, J.F.T. Yeast genetics: a manual of methods/J.F.T. Spencer, D.M. Spencer, I.J. Bruce. p. cm. Includes index.
ISBN 0-387-18805-3 (U.S.)
1. Yeast fungi-Genetics-Laboratory manuals. 2. Fungi-Genetics-Laboratory manuals. I. Spencer, Dorothy M. II. Bruce, I.J. III. Title. QK617.5.S64 1988 589.2'330415-dc 19 88-20066

This work is subject to copyright. All rights are reserved, whether the whole or part of the material is concerned, specifically the rights of translation, reprinting, re-use of illustrations, recitation, broadcasting, reproduction on microfilms or in other ways, and storage in data banks. Duplication of this publication or parts thereof is only permitted under the provisions of the German Copyright Law of September 9, 1965, in its version of June 24, 1985, and a copyright fee must always be paid. Violations fall under the prosecution act of the German Copyright Law.

© Springer-Verlag Berlin Heidelberg 1989
Printed in Germany

The use of registered names, trademarks, etc. in this publication does not imply, even in the absence of a specific statement, that such names are exempt from the relevant protective laws and regulations and therefore free for general use.

Typesetting, printing and binding: Brühlsche Universitätsdruckerei, Giessen
2131/3130-543210 – Printed on acid-free paper

Contents

Introduction 1

Part I. "Classical" Yeast Genetics 3
1. Mating 4
   a) Drop-Overlay Method 4
   b) Mating of Prototrophic Haploid Strains 5
2. Sporulation 8
   a) Spore Isolation 10
      i) Heat Killing of Vegetative Cells 10
      ii) Ether Killing of Vegetative Cells 11
      iii) Sporulation of Protoplasts and Bursting of Vegetative Cells . . 12
      iv) Separation of Asci and Vegetative Cells in Biphasic Systems . . 13
      v) Separation of Asci and Vegetative Cells on Renografin
         Density Gradients 14
3. Tetrad Analysis 16
   a) Operation of Micromanipulators 16
   b) Variants: Spore-Spore or Spore-Cell Pairing 21
   c) Detection of Post-Meiotic Segregation 22
      i) Procedure 1: Dissection on the Surface of an Open Plate 22
      ii) Procedure 2: Dissection on an Inverted Plate
         (Fogel's Procedure) 23
4. Enzymes for Digestion of Yeast Ascus Walls 25
5. Ancillary Methods for Handling Clones Obtained by
   Ascus Dissection 26
   a) Replica Plating 26
      i) Velvet Pad 26
      ii) Filter Paper 27
   b) Mating Type Determination by Cross-Stamping 28
6. Mutagenesis 30
   a) Ultraviolet and X-Irradiation 30
      i) Procedure 1: (Liquid Medium) 30
      ii) Procedure 2: (Solid Medium) 31

	iii) General Note on Mutagenesis	32
	iv) Special Note: Induction of Mitotic Recombination by Ultraviolet Irradiation	33
b)	Drying	33
	i) Freeze-Drying	34
	ii) Vacuum Drying	34
c)	Chemical Mutagens	35
	i) Nuclear Mutations	36
	ii) Mitochondrial Mutations	36
d)	Isolation of Particular Mutants or Classes of Mutants	39
	i) Mutator Mutants	39
	ii) Temperature-Sensitive Mutants	42
	iii) Isolation of *kar* Mutants (Defective in Nuclear Fusion)	42
	iv) Cell Wall Mutants	43
	v) Antibiotic-Sensitive Mutants. "Kamikaze" Strains	47
	vi) PEP4 Mutants	48
	vii) Membranes. Fatty Acid and Inositol-Requiring Mutants	49
	viii) Mutants Auxotrophic for 2'-Deoxythymidine 5'-Monophosphate	52
	ix) Glycolytic Cycle Mutants	53
	x) Secretory Mutants, Transport Mutants, etc.	57
7. Mapping and Fusion		58
a)	Tetrad Analysis	58
b)	Determination of Centromere Linkage	59
c)	Assignment of a Gene to a Particular Chromosome: Trisomic Analysis	60
	i) Known Chromosome is Disomic	60
	ii) Chromosome Bearing the Unmapped Gene is Disomic	60
	iii) Multiply Disomic Strains	60
	iv) Super-Triploid Method	60
d)	Mitotic Mapping Techniques	61
	i) Mitotic Crossing-Over	61
	ii) Mitotic Chromosome Loss	61
e)	Mapping by Chromosome Transfer	62
f)	Fine-Structure (Intragenic) Mapping	62
g)	Strategies for Mapping According to the Above Methods	63
8. Protoplast Formation and Fusion		64
a)	Regeneration in Solid Medium	64
b)	Regeneration in Liquid Medium	67

Part II. Methods Using Direct Manipulation of DNA and RNA 71
1. Separation of Large DNA Molecules, Greater than 25 Kbp, by
   Gel Electrophoresis 71
   a) Preparation of Intact Chromosomal-Sized Yeast DNA Molecules . 75
   b) Gel Preparation 77
   c) Size Standards 78
   d) Restriction Digests 80
   e) Conditions for Electrophoresis 81
      i) Loading the Gel 81
      ii) Running the Gel 81
   f) OFAGE 82
   g) FIGE 82
2. Isolation of Pure High-Molecular-Weight DNA from the Yeast
   Saccharomyces cerevisiae 85
   a) Protoplasting 85
   b) Protoplast Lysis and DNA Recovery 86
3. Transformation of Yeast: *Saccharomyces cerevisiae* 88
   a) Protoplast (Spheroplast) Transformation 89
   b) Intact Cell Transformation 90
4. Plasmid Isolation from Yeast 91
5. Rapid Isolation of DNA from Yeast 93
6. RNA Isolation 94
   a) With Glass Beads 95
   b) Without Glass Beads 96

Subject Index 98

Introduction

The manual consists of two main sections. The first includes the essential, sometimes laborious, procedures for handling yeasts, for inducing mating and isolation of hybrids, for inducing sporulation and isolation of single-spore clones, with some details of tetrad analysis, and including techniques and ancillary equipment for use of the micromanipulator. There are also procedures for induction of mutants by physical and chemical agents, and for isolation of particular types of mutants, such as to temperature sensitivity, for increased frequency of mutations, for mutations in the mitochondrial genome, both to the petite colonie form and to resistance to antibiotics, for mutations in that part of the yeast genome controlling the glycolytic cycle, and numerous others. Mapping of mutations is discussed briefly, though this aspect of yeast genetics is probably one which should not be undertaken until the investigator has gained a certain amount of experience in the field. However, as is pointed out in the pertinent part of the manual, the task of mapping has been tremendously simplified by the availability from the Yeast Genetics Stock Center at the University of California at Berkeley of a set of auxotrophic strains designed to permit mapping of most unknown genes with a minimum number of crosses and tetrad analyses. The first section concludes with the description of methods for hybridization of yeasts by protoplast fusion, which has been described as the poor man's system for genetic engineering. Be that as it may, the procedure has a great many applications in the construction of improved industrial yeast strains, both by fusion of strains of the same species, usually but not necessarily *Saccharomyces cerevisiae*, and for introduction of desirable genes from other species from different genera, by gene transfer associated with cytoduction and cybrid formation.

The second, and very important section, deals with the more up-market methods used in yeast genetics and genetic engineering by recombinant DNA techniques. In particular, Dr. Bruce emphasizes the use of pulsed-field gel electrophoresis, including OFAGE, the grand old procedure of them all, FIGE, and CHEF, the most sophisticated of the three procedures, which achieves sharper separations of a range of sizes of high-molecular-weight

DNA molecules by contour clamping of individual electrodes in arrays. Of the three systems, FIGE is probably the simplest for the beginner to use, and it also gives excellent separations of large molecules of DNA, the size range separated being variable by simple changes in the pulse frequency and timing. Also included are techniques for isolation and analysis of high molecular weight DNA and RNA and their characterization.

Recombinant DNA techniques, which permit isolation of individual genes and their re-insertion into the yeast genome at different sites, fusion of genes with promoters, leader sequences, fusion with genes from other organisms (lacZ gene, for example), any many other such operations, have greatly enhanced the range of problems which can be solved by the yeast geneticist. While it is not possible to discuss all of the possible uses of gene cloning and related recombinant DNA techniques, we have described some of the more important of the basic methods, which will enable the yeast geneticist to go on to the solution of many problems, which could not even be formulated, let alone attempted, before the advent of these methods.

The manual assumes a basic knowledge of microbiological and biochemical techniques, but given this, our hope is that it will enable both the serious student and the experienced investigator to gain a wider knowledge of the current methods in use in this field, and to accelerate the pace of their investigations.

Part I. "Classical" Yeast Genetics

This stage began, essentially, with the discovery of haploid and diploid phases in the life cycle of *Saccharomyces cerevisiae* by Winge (1935), and of Mendelian segregation in *Saccharomycodes ludwigii* (Winge and Lautsten 1939) and the isolation of heterozygous strains in which haploid clones, obtained by dissection of yeast asci and separation of the individual spores, did not self-diploidize and could be used to re-form the diploid state by mating of vegetative cells of opposite mating types, or by pairing of spores or of spores and vegetative haploid cells by micromanipulation. Auxotrophic mutants of the haploid clones can be obtained by mutagenesis with X-rays, UV-irradiation, chemical mutagens or by other methods. Complementing mutations in clones of opposite mating types can be used to obtain prototrophic diploids, and these can be sporulated and haploid clones can be isolated and used to demonstrate recombination of characters during meiosis, linkage of characters, and the construction of maps of the yeast chromosomes. These techniques have been most widely used in the study of the genetics of *Saccharomyces cerevisiae*, but have also been extremely useful in the study of the genetics of other yeast species, including *Schizosaccharomyces pombe*, *Yarrowia lipolytica* (a species of importance for its ability to metabolize alkanes and lipids, and to produce extracellular proteases), *Kluyveromyces lactis*, *Pichia pinus*, *Hansenula polymorpha* and other methylotrophic yeasts, and various other species. Some years ago, mating types and a sexual cycle were observed in the basidiomycetous species, *Rhodosporidium toruloides* (imperfect stage *Rhodotorula* sp.) and *Leucosporidium scottii*, and in a group of ascomycetous yeasts forming needle-shaped spores, of the genus *Metschnikowia*.

It might be thought that the discovery of restriction enzymes, cloning of genes, and transformation of yeasts with various plasmid-like vectors, would have by now superseded the use of the methods of classical genetics in the production of new yeast strains and in characterization of known strains, but in fact, the one group of techniques usually complements the other, one taking up where the other leaves off, so that a knowledge of the methods of "classical" yeast genetics as well as of those based on recombinant DNA techniques is highly desirable, especially for the yeast geneticist who desires to construct improved strains of industrial yeasts.

REFERENCES

Winge O (1935) Cr Trav Carlsberg 21:77–112
Winge O, Lautstsen O (1935) Cr Trav Carlsberg 22:357–374

1. Mating

Mating of haploid yeast strains can be done in a number of ways, either by mixing the yeast strains on a rich medium and isolating the resulting zygotes by micromanipulation, or by some variant of the drop-overlay method. The latter is frequently used for mating auxotrophic strains having complementary requirements. For prototrophic parental strains, of course, hybrids can only be isolated by micromanipulation of the zygotes from the mating mixture.

a) Drop-Overlay Method

Required

1. 24-h cultures of auxotrophic haploid strains (opposite matings types, having complementary nutritional requirements), to be mated.
2. Plates of minimal medium containing a carbon source (yeast-nitrogen base-glucose (2%), for instance).
3. Sterile water blanks, 9.0 and 9.9 ml, sterile pipettes, Pasteur pipettes and other glassware.

Procedure

1. Dilute the culture approximately 1:10 with sterile water and wash once or twice with sterile water and make back to original volume.
2. With a Pasteur pipette, put 2 drops of the suspension of the first strain on a plate of minimal medium (as above), in well-separated locations. Let dry.
3. Put a drop of the second suspension on top of one of the dried drops of the first, and another drop in a new location. Let dry. There should be one drop containing cells of both strains, and two other drops of one of each of the other strains, as controls (Fig. 1).

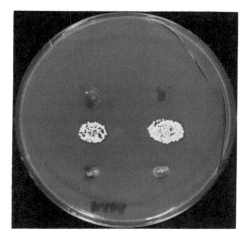

Fig. 1. Mating of haploid strains. Drop overlay pattern

4. Incubate 3–4 days. Microcolonies should occur in profusion if the strains are active maters, in the drop containing both strains. Occasional microcolonies may arise in the controls if only a single auxotrophic marker is present, brought about by reversion of the marker. These are not normally observed where two or more markers are present.
5. Restreak the prototrophic diploids on minimal medium (Fig. 2). Test for sporulation. Haploid revertants do not sporulate.

b) Mating of Prototrophic Haploid Strains

This can be done in the same way as for auxotrophic haploids, though isolation of the resulting diploids is more laborious, as the complementation method on minimal medium cannot be used. Two methods at least are possible:

1. Isolation of diploids by streaking out the mating mixture and random isolation of colonies, which are then tested for the ability to sporulate. This method can be varied by selecting and isolating a number of large cells by micromanipulation, followed by testing the strains for sporulation as before. This method has the advantage of ensuring that the strains arise from single cells. If homothallic strains are to be mated, spore-spore or spore-cell pairing (using haploid mating strains for the vegetative cells) by micromanipulation can be used. The difficulty in this case is to determine whether the resulting colonies are actually a cross, unless some marker is introduced in the haploid strain, which can be detected later in the hybrid. Markers such as are used in the "rare-mating" method are probably the most satisfactory.

Fig. 2. Patterns for restreaking yeast colonies

2. Possibly the quickest method is to isolate zygotes from the mating mixture. These appear, usually from 2–6 h after mixing of the strains on a complete medium, though in some cases at least, zygotes can be observed in the mating mixture for 24–48 h after mating begins. Individual zygotes can be isolated from the mixture with a micromanipulator, though zygotes are not as readily picked up as tetrads or individual spores (see Section "Micromanipulation").

3. "Rare-mating" techniques. This method is based on the observation by Gunge and Nakatomi (1971) that mating-type switching occurs at low frequencies in *Saccharomyces cerevisiae*, in diploid cells or cells of higher ploidy, and leading to the appearance of mating cells, which will mate and form hybrids if cells of another mating type are present. The method was adapted by us for use in obtaining hybrids from industrial yeast strains, which often sporulate poorly or not at all, and which generally form nonviable spores if any are produced (Spencer and Spencer 1977). The diploid (or polyploid) strain is converted to the petite form by any one of numerous methods, including treatment with acriflavin or ethidium bromide, by starvation, or simply spreading an appropriate dilution of the yeast culture on a plate of normal medium, incubating for a few days, and selecting small colonies, which are then tested for inability to grow on non-fermentable substrates (glycerol or lactate). The petite strain is then mated with an auxotrophic mating strain which may be haploid or diploid, of either mating type, as in most cases there will be switching to either mating type in the petite. Hybrids are isolated on selective medium, as follows:

Required

1. Petite mutants of the industrial strain (see petite mutants).
2. Mating strain of laboratory yeast, carrying any auxotrophic markers desired.
3. Culture media: 1. Yeast extract-glucose broth.
   2. Yeast-nitrogen base (YNB)-glycerol agar, containing 3% ethanol to inhibit sporulation in the hybrids.
4. Sterile pipettes, centrifuge tubes, and possibly sterile membrane (Millipore) filters; sterile McCartney bottles, if this variant of the method of mating is to be used.

Procedure

1. Grow cultures, 24–48 h in YEP-glucose broth. Usually enough culture can be grown in 5 ml of medium, aerated on a roller drum or rotary shaker held in an inclined rack.
2. Either: mix cultures in YEP broth and incubate in still culture in shallow layers to allow aeration, for 5–6 days,

or, mix cultures and filter on a membrane filter, to pack the cells tightly. Make sure the cells do not dry out.

Or: mix the cultures and centrifuge to pack the cells tightly.

3. After allowing sufficient time for mating to take place (a few hours for the two latter variants), recover and wash cells twice in sterile water or buffer.
4. Spread a heavy suspension of the cells on the YNB-glycerol-ethanol agar, and incubate until colonies appear. The auxotrophic laboratory strain will not grow on minimal medium, and the petite of the industrial strain will not grow on a non-fermentable carbon source.
5. Restreak the colonies of respiratory-competent prototrophs which appear, on YNB-glycerol-ethanol agar. Isolate and store.

Note

Homothallic strains may be held in the haploid state by dissecting the asci directly on to a slab of acetate agar (McClary's medium), and using the small colonies arising from the spores in mass mating with known laboratory haploids, or in pairing by micromanipulation with haploid cells or spores of other strains. If the strains are prototrophic, then the usual system of isolating zygotes directly by micromanipulation must be used. If some kind of marker is naturally present, or can be introduced into the strains involved, identification of the hybrids is simplified (see Palleroni 1961).

REFERENCES

Gunge N, Nakatomi N (1971) Genetics 70:41–58
Spencer JFT, Spencer Dorothy M (1977) J Inst Brewing 83:287–289
Palleroni N (1961) Phyton, Buenos Aires 16:117–128

2. Sporulation

Diploid hybrids (or hybrids of higher ploidy) having been obtained, sporulation can then be induced in a number of ways, mostly involving growth of the culture on a medium containing potassium acetate, glycerol or other non-fermentable substrate. Sporulation may also be induced in raffinose-acetate solutions. The presence of K ions improves sporulation. For most yeast strains, McClary's medium (Fowell 1969) gives adequate sporulation. Growth on a relatively rich pre-sporulation medium such as yeast extract-peptone glucose medium is desirable. Sporulation may be induced in liquid or on solid medium, or in protoplasts as well as in intact cells. Induction of sporulation in protoplasts has the advantage that the unsporulated protoplasts may be burst by addition of water to the osmotically stabilized sporulation medium which must be used in this method, and the spores can either be isolated by micromanipulation of the naked tetrads or harvested, free of unsporulated vegetative cells, and subjected to random spore analysis.

Required

1. Presporulation medium. Yeast extract-peptone-glucose medium, liquid or solidified with agar.
2. McClary's sporulation medium:

Potassium acetate	1 g
Yeast extract	0.25 g
Glucose	0.1 g
Agar (if desired)	1.5 g
Water	100 ml

   If this medium is used in liquid form, it must be aerated adequately for good sporulation.

Procedure

1. Inoculate the yeast strains on to YEP-glucose medium and incubate for 48-72 h at 25-30°. If sporulation in liquid medium is required, inoculate 4 ml of the above medium in a 16-mm culture tube and incubate with shaking for the same time and temperature.
2. Inoculate a plate or tube of McClary's medium from the presporulation medium (for solid medium, streak on the medium, for liquid medium, inoculate from the broth culture at a ratio of 1:10. Two ml of sporulation medium in a 16 mm culture tube, with 0.2 ml of presporulation culture, for instance).
3. For sporulation on solid medium, incubate at 20-30° for up to 7-10 days. Cultures which sporulate readily will generally show the first asci in 48 to 72 h. Some cultures sporulate better at lower temperatures, in which case, it may be necessary to incubate for a week or more.
4. Cultures sporulated in liquid cultures should be incubated on a shaker operated at a high enough speed to give good aeration. Sporulation usually occurs in 3-4 days. Confirm sporulation, in both methods, by microscopic observation.
5. Centrifuge cultures sporulated in liquid medium, and wash cells and asci once or twice in sterile water, and store in the refrigerator. If the tubes are sealed with parafilm to prevent drying, the spores will remain viable, normally, for several weeks.
6. Cultures sporulated on solid medium may be stored successfully for a week or two at refrigerator temperature, if kept from drying.

Notes

1. Better sporulation and spore viability can be obtained in some strains by incubating at 18-20° instead of at 25-30°.
2. Some strains may be sporulated adequately in raffinose-acetate medium (2% + 1%), or on media containing glycerol as sole carbon source.
3. Strains of *Saccharomyces diastaticus*, and some of our intergeneric fusion hybrids, having *S. diastaticus* as a parent, will often sporulate well on yeast extract-soluble starch agar.
   Sometimes yeasts sporulated on media other than those containing acetate form an ascus wall which is very difficult to dissolve enzymatically.
4. *Schizosaccharomyces pombe* will often sporulate adequately on malt extract agar. In this species, the ascus wall dissolves without the use of enzymatic treatment, which makes the isolation of the ascospores less complicated.
5. The above procedures using acetate medium for sporulation, apply normally to *Saccharomyces cerevisiae* strains. Other yeast species (*Yarrowia lipolytica, Saccharomycopsis fibuligera, Saccharomycopsis capsularis, Hansenula* and *Pichia* species, *Debaryomyces* species, and various others) can be sporulated readily on malt extract agar, Gorodkowa agar and other specialized media (Lodder 1970, 1984).

a) Spore Isolation

Random Spore Isolation. In these techniques, some way of eliminating the unsporulated vegetative cells is essential. This can be done either by killing the vegetative cells, or by separating them physically from the spores in some way.

i) Heat Killing of Vegetative Cells

This method is based on the slightly greater heat resistance of spores as compared to vegetative cells. However, different strains differ in the resistance of the vegetative cells and spores to heat, so that the actual times and temperatures used should be determined for each strain used. The temperature will normally be between 54 and 60 °C.

Required

1. Yeast cells, sporulated as above.
2. Sterile water and culture tubes.
3. Water bath, set at the predetermined temperature.

Procedure

Suspend the sporulated yeast culture in sterile water, and hold in the water bath at the desired temperature (example, 59°), as determined previously, for 10-15 min. Plate the treated spore suspension on malt agar or YEP-glucose agar.

Notes

1. Industrial yeast strains, if polyploid or aneuploid, may form spores of varying ploidy, so that diploid strains as well as haploids, which generally form smaller colonies, may be isolated. The isolates should be tested for the ability to sporulate as well as for mating type.
2. Industrial yeast strains are frequently homothallic, so that single spore clones which were originally haploid may well be diploid when isolated.
3. Strains forming few viables spores will often form asci with no more than one viable spore each, which facilitates the isolation of single-spore clones by this and the following method. It is necessary to verify the purity of the clones if possible.

ii) Ether Killing of Vegetative Cells (Dawes and Hardie 1974)

This method is based on the differential sensitivity of the vegetative cells and spores to exposure to diethyl ether.

Required

1. Sporulated yeast culture.
2. Sterile water.
3. Diethyl ether.
4. McCartney bottles or roller tubes.

Procedure

1. Make a suspension of the yeast culture in water.
2. Place 5 ml of suspension in roller tubes or McCartney bottles, cool to 4 °C, add an equal volume of ether, close tube tightly and place on roller apparatus in cold room, or otherwise agitate for 10 min.
3. Separate ether and water layers, and remove ether from water layers under vacuum.
4. Plate out suspension on malt agar or YEP glucose agar.

Notes as for heat killing also apply to ether killing.

REFERENCE

Dawes IW, Hardie ID (1974) Selective killing of vegetative cells in sporulated yeast cultures by exposure to diethyl ether. Molec Gen Genet 131:281–289

iii) Sporulation of Protoplasts and Bursting of Vegetative Cells

Strains which form protoplasts readily can be grown on presporulation medium, converted to protoplasts (see section on protoplast formation) and the protoplast suspension sporulated, after which the protoplasts can be burst, giving a relatively pure spore suspension.

Required

1. Yeast culture, 16–24-h-old (exponential growth phase), in YEP-glucose medium.
2. Pretreatment and protoplasting solutions, as in Section 8, (Protoplast Fusion), 0.6 M KCl solution (all solutions sterile).
3. Snail enzyme, Zymolyase, Novozyme or other enzyme suitable for digesting yeast cell walls.
4. Osmotically stabilized sporulation medium.
5. Sterile water, culture tubes, pipettes, etc.
6. Ultrasonicating apparatus or Potter-Elvehjem homogenizer with Teflon pestle (sterile).

Procedure

1. Grow yeasts to exponential phase and convert to protoplasts (Section 8, Protoplast Fusion).
2. Transfer protoplasts to osmotically stabilized sporulation medium (liquid) and incubate (shallow layers for better aeration) until spore formation has taken place.
3. Recover sporulated cells by centrifugation if necessary and add sterile water to burst unsporulated vegetative cells.
4. Lightly sonicate the suspension, or give 1 or 2 strokes in the Potter-Elvehjem homogenizer, to separate the spores.
5. Plate on malt agar or YEP-glucose agar. Various substances (15% gelatin solution, etc.), may be added to aid in keeping the spores separated during plating.

iv) Separation of Asci and Vegetative Cells in Biphasic Systems (mineral oil-water, dextran-polyethylene glycol), or on density gradients (renografin, urografin, etc.)

1. Separation of spores from vegetative cells by partition in mineral oil. This method is based on the lipophilic properties of the spore wall. The spores may be released from the asci by grinding with powdered glass, or by treatment with enzymes as above. The spores obtained by grinding and shaking with mineral oil often clump severely, but may be separated by addition of 15% gelatin solution. Spores released from the asci by enzymatic treatment may be separated by sonication.

Required

1. Sporulated cultures.
2. Sterile mineral oil.
3. Potter-Elvehjem homogenizer with Teflon pestle or similar grinding apparatus *or* snail enzyme, mushroom enzyme (see section 4, Enzymes for removal of cell walls, for method of preparation), Novozyme, Zymolyase, or other wall-dissolving enzyme preparation.
4. Gelatin, if required.
5. Sterile water.
6. Malt agar or YEP-glucose agar plates.
7. Other glassware as required.

Procedure

1. Mix a water suspension of a sporulated culture with powdered glass.
2. Place in homogenizer and grind for 10-15 min or as necessary, cooling as required. Most of the vegetative cells are usually broken at this stage.
3. Shake the suspension with sterile mineral oil, or give the oil-suspension mixture a few more strokes with the homogenizer pestle.
4. Transfer the suspension to a sterile centrifuge tube and separate the oil and water layers by normal centrifugation. The spores should be retained in the oil layer; the vegetative cells and debris should be pelletted in the water layer.
5. Separate the layers with a Pasteur pipette and wash the oil layer with water, as many times as are necessary to remove any remaining vegetative cells.
6. Observe the spore suspension under the microscope. If the spores are clumped, add 15% gelatin solution and grind again lightly, without glass powder.
7. Spread or streak the diluted suspension on malt agar or YEP-glucose plates.
8. Isolate colonies and test for sporulation and mating. Single-spore clones should be maters, unless the original strain was homothallic.

Or, if enzymatic treatment is used:

1. Suspend sporulated culture in water (1-2 ml depending on the amount of culture available) and add enzyme. For snail enzyme, the dilution may be from 1:4 to 1:40, depending on the purity of the enzyme preparation.
2. Incubate until spores are released from asci.
3. Wash suspension once or twice with water.
4. Add mineral oil and shake thoroughly or give a few strokes of the pestle in the homogenizer.
5. Centrifuge emulsion to separate layers. Spores should be in the oil layer.
6. Repeat washing of oil layer if required, until there are no vegetative cells observable in the oil layer.
7. Sonicate suspension lightly to break up clumps. Gelatin solution may be added if desired.
8. Plate on malt agar or YEP-glucose agar.
9. Isolate colonies and test as in previous protocol.

Notes

1. Heat treatment as described previously may be used to eliminate any remaining vegetative cells in the spore suspension.
2. Dextran-polyethylene glycol partition.

v) Separation of Asci and Vegetative Cells on Renografin Density Gradients

Required

1. Sporulated culture.
2. Sonication apparatus.
3. 20, 35 and 43% renografin solutions.
4. Snail enzyme.
5. Sterile centrifuge tubes, 15 or 25 ml.
6. Sterile pipettes, Pasteur pipettes.

Procedure

1. Sonicate sporulated culture lightly to break up clumps.
2. Spin down sporulated culture (approximately 10 ml).
3. Resuspend in 1 ml of 35% renografin.
4. Construct a gradient of 3 ml of 35% renografin layered on 3 ml of 43% renografin.
5. Layer the spore suspension in renografin on top of the gradient, using a Pasteur pipette.
6. Layer 3 ml of 20% renografin on top of the spore layer.

7. Centrifuge at 2500 rpm (approximately) for 15 min. The vegetative cells will be visible at the interface between the 35% and 43% renografin layers and can be removed with a Pasteur pipette and discarded.
8. The pellet should contain only asci. If there are still vegetative cells present in the pellet, re-run the material on a single-step gradient of 35% and 43% renografin solutions, constructed as before.
9. Wash pellet if desired.
10. Treat asci with snail enzyme (or other) to liberate spores, plate and isolate as for other methods of random spore isolation. Sonicate lightly to separate spores. Otherwise, isolate single-spore clones by micromanipulation.

Notes

1. The renografin concentration at the bottom of the gradient may have to be varied for adequate separation of asci and vegetative cells in different yeast strains.
2. Cells which will eventually sporulate can be separated from vegetative cells which will not sporulate, on a similar gradient, as follows: The yeast culture is grown to stationary phase in YEP-glucose broth (approximately 48 h), and the cells are recovered, washed, sonicated lightly and resuspended in 4 ml of 20% renografin solution. This suspension (2 ml) is layered over 10 ml of 35% renografin solution in a 30 ml chlorex centrifuge tube, and centrifuged as above. The cells at the interface are recovered with a Pasteur pipette (Fraction I). These cells are large, vacuolated, and will sporulate on the appropriate medium. The pellet at the bottom of the tube (Fraction II) contains the vegetative cells, which are smaller, unvacuolated, and will not sporulate. If a clear separation is not obtained at first, repeat the procedure. The Fraction I cells are washed and placed in liquid sporulation medium or spread on solid sporulation agar and incubated.
3. Subpopulations from cultures in the early stages of sporulation can be separated and recovered in a similar manner using linear Urografin density gradients (in 16.5 ml centrifuge tubes), 1.13 to 1.22 g cm^{-3}, the optimum range being 0.06 g cm^{-3}. At least 8 fractions were observed (Dawes et al. 1980).

REFERENCE

Dawes IW, Wright JF, Vezinhet F, Ajam N (1980) J Gen Microbiol 119:165–171

3. Tetrad Analysis

a) Operation of Micromanipulators

There are several makes of micromanipulators available commercially, all of which can be used for dissection of yeast asci and isolation of the ascospores. Most of these are designed as free-standing instruments, but in at least one, the needle-holder and adjusting screws, both for the operation of the needle itself and for moving the dissecting chamber, are mounted on the microscope stage itself. One of the earliest commercial instruments produced was the de Fonbrune micromanipulator, in which the controls were combined into one "joystick" arrangement, which operated a set of hydraulic plungers connected by flexible lines to aneroid capsules in the needle holder. In other instruments, the movement is purely mechanical. Those instruments which are designed for microinjection of animal cells and plant protoplasts, and which are capable of rather slow, but very precise movements, can be used for dissection of yeast asci, but are possibly less easy to manipulate than those instruments specifically designed for the isolation of yeast ascospores.

Micromanipulation and isolation of yeast cells and ascospores in this way is often regarded as an arcane skill, akin to the practice of the black arts. However, dissection of yeast asci is a purely mechanical skill which can in fact be learned relatively easily and quickly by any research worker. This section is designed to shorten the time required for learning the skill and to enable the beginner to start using the technique in investigations involving yeast genetics.

Required

1. Micromanipulator, of any brand available (Fig. 3). (Sherman 1973, has described a relatively simple and inexpensive instrument which can be assembled from commercially available components).
2. Microscope, with a stage having a slide holder capable of holding the dissection chamber (described below), and having a reasonable range of movement. It should preferably have 20x eyepieces, but dissection using 10x eyepieces is perfectly possible. The operator may feel eyestrain somewhat sooner.
3. Dissection chamber. The chamber can be assembled by sticking a metal strip (any metal flexible enough to be bent into a U-shape), approximately $7'' \times 3/8''$, bent to fit, to a 37×75-mm microscope slide with expoxy cement. Smaller or larger slides can be used, but working is more tedious using a smaller chamber, and one assembled on a larger slide may be difficult to fit into the slide holder (Fig. 4).

Fig. 3. Singer MK111 Micromanipulator in a typical set up for ascus dissection from below the plate. The Petri dish is raised on a plastic mount and the disposable glass microtool, also made by Singer instruments, is held in a special holder beneath the plate

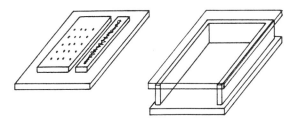

Fig. 4. Needles, agar slab and dissecting chamber

4. Other slides, 37 × 75 mm, to hold dissecting slab and cover chamber.
5. Alcohol, for sterilizing slides and other implements. Spare slides can be kept in a screw-top jar of alcohol until needed.
6. Tweezers (forceps), microspatulas, small loop (0.5–1 mm diameter).
7. Flat-ended dissecting needles (Fig. 4). These can be drawn from borosilicate (Pyrex) glass rod, 2 or 3 mm diameter. Soda glass has different melting characteristics and is usually not satisfactory. The needles are pulled in two stages, first to about 0.2 mm diameter, in an ordinary flame, and then to the desired degree of fineness in a very small flame, such as the pilot flame of the burner, if it can be turned down far enough, or in a microburner made by clamping a hypodermic needle in a holder and connecting it to the gas supply. The final section of the needle should be approximately visible to the naked eye. Finally, the fine section is bent at right angles to the shaft, in the small flame. It may prove easier to pull the needle to its final thickness in a third stage, after bending at right angles. The needle is then cut off to the desired length (6–7 mm) with a razor blade or very sharp knife. If the needle proves to have a spike at one side, it may be retrimmed.

    The intermediate stage should be long enough to be slightly flexible (3–4 cm).

    For the de Fonbrune micromanipulator, a smaller rod, about 1 mm diameter is necessary, to fit in the needle holder.
8. Dissecting agar. Plates containing about 10 ml of 2–2.5% agar, plus 1% yeast extract and 2% glucose, are satisfactory. Dissecting agar may be made up in 10 ml batches and kept in closed tubes or McCartney bottles until needed.
9. Plates of YEP-glucose agar.
10. Sporulated yeast culture.
11. Enzyme for dissolving the ascus walls. Snail enzyme, zymolyase, mushroom enzyme or other preparation may be used. It is desirable to centrifuge some of the preparations before use.
12. Sterile tubes or screw-capped bottles (small).
13. Template, for cutting slabs of dissecting agar, approximately 2 × 3.5 cm, from the plate.

Procedure

1. Dilute the enzyme as required. This varies considerably with the source of the enzyme, but generally ranges from 1:4 to 1:40. Some yeast strains produce asci which are more resistant to enzymatic removal than others.
2. Suspend enough for the sporulated yeast culture in the diluted enzyme to give a slightly turbid suspension. If the volume of enzyme used is ap-

proximately 0.2 ml, only a trace of culture will be needed, and this will yield many times the number of asci required for normal purposes.
3. Incubate the suspension at 35 °C for 15-30 min, or until the ascus walls have been dissolved sufficiently to allow release of the spores.
4. Take a 1" × 3" slide from the alcohol jar, flame and place it on the dissecting chamber.
5. Place the Petri dish of dissection agar on a template, and using a microspatula dipped in alcohol and flamed, cut a slab of dissection agar approximately 2 cm × 5 cm. Place the slab on the slide which has previously been flamed and placed on the dissection chamber.
6. Cut a thin section (approximately 2 mm wide) off one side of the slab. Leave on the slide, next to the main slab (see diagram, Fig. 4).
7. With the small loop (0.5-1 mm), streak a loopful of the digested spore suspension along one edge of the narrow strip. Leave the other edge of this strip clear for moving cells and asci around without contaminating the main slab.
8. Invert the slide and slab over the dissecting chamber, and place the chamber in the slide holder of the microscope, under the low-power (10x) objective. Focus on the bottom of the slab, in the area of the spore suspension.
9. Mount the dissecting needle in the micromanipulator, and position it at the correct height below the slab. Locate the tip of the needle in the center of the microscope field using the coarse controls. The needle tip is often easier to locate if the microscope is focused on one corner of the narrow strip carrying the spore suspension.
10. With the stage controls, move the chamber to locate an ascus which is reasonably well separated from the vegetative cells and other asci. There will be plenty of time to indulge one's feelings of superiority over the instrument by picking out asci from crowded fields later, when you have mastered the technique more fully.
11. With the micromanipulator controls, raise the tip of the needle to just touch the surface of the agar close beside the tetrad of spores. Lower the needle slightly to lift the tetrad off the agar. If the tetrad fails to adhere to the needle, repeat the procedure until the tetrad comes away from the slab and remains on the needle. (The appearance of the needle in the microscope field will come as a surprise. It will appear many times the size of the spores, which disconcerts some beginning operators. The needle is designed, not to push the spores around, but to provide a suitable surface for the spores to adhere to by surface tension. Sometimes an even larger needle may be more useful for picking up vegetative cells, which are in their turn considerably larger than spores).

12. With the stage controls, move the chamber and slab to a location so that the spores can be set down on the main slab, approximately 3 mm from the edge. With the micromanipulator control, touch the tip of the needle to the agar and lift it off again. The tetrad of spores should remain on the agar. If the spores remain stuck to the needle, touch the agar with the tip and try again, until success crowns your efforts.
13. Again touch the agar lightly with the tip of the needle, beside the tetrad. Tap the bench top or the base of the micromanipulator, just hard enough to make the needle vibrate slightly, until the spores separate.
14. Pick up the spores as before, leaving one or more behind. Move the dissecting chamber with the transverse stage control, about 2 mm. Adjust the microscope focus if necessary. Touch the agar surface with the needle tip again, and set down one or more spores. Pick up all but one spore, move the chamber again, and repeat until all four spores have been set down in a line across the slab. Move the needle back to the top edge of the slab and make a mark on the agar to relocate the spores if necessary.
15. With the other control, move the chamber 3 mm along the stage, and again mark the edge of the agar.
16. Move the chamber to bring the spore suspension into the field of the microscope again. Locate another ascus and pick up and transfer as before.
17. Move the chamber back to the mark, then position the chamber to set down the second tetrad, which is then dissected as before. Usually 10 to 12 tetrads can be dissected on one slab.
18. Move the micromanipulator to one side, or otherwise remove the needle from the area. (It is possible to remove the top slide from the chamber without disturbing the micromanipulator or damaging the needle, but this is not part of the learner's procedure). Remove the chamber and top slide from the stage, set down and again invert the slide so that the slab with the spores is on top.
19. Flame the microspatula in alcohol again and lift off the main slab and spores, and place it on the surface of a plate of YEP-glucose agar. Discard the small slab with the remaining asci and cells and place the slide in the alcohol jar.
20. Incubate the slab on the YEP-glucose agar at 30 °C until colonies appear (2-3 days, normally). Pick the colonies to a plate of YEP-glucose agar, using sterile toothpicks, and a template drawn so that 32 single-spore clones can be placed on one Petri dish.
21. Incubate again at 30 °C and replica plate the streaks to plates of selective media, if desired.

b) Variants: Spore-Spore or Spore-Cell Pairing

Required

1. Sporulated culture or cultures (if spore-spore pairing is desired), treated with enzyme as before, to dissolve ascus walls.
2. Fresh culture of mating strain desired.
3. Other equipment and materials for micromanipulation of yeast cells and spores.

Procedure

1. Set up dissecting chamber, with flamed upper slide as before.
2. Cut slab slightly wider than for simple ascus dissection, and place on sterile slide on chamber.
3. Cut two narrow strips, as before, one from each side of slab.
4. Streak spore suspension on one strip.
5. Dissect asci from the first culture and set out spores on slab, but mark the location of each spore as well as marking the edge of the slab.
6. When the dissections are complete, remove the micromanipulator and needle, remove the dissecting chamber from the microscope stage, invert the slide again, and, using a flamed spatula, remove the narrow strip carrying the spore suspension.
7. Reflame the spatula, and move the other narrow strip to the edge with the markings, next to the main slab.
8. Streak a suspension of the mating strain (vegetative cells) or the second spore suspension, if desired, on the strip, invert the slide over the chamber again, replace the chamber on the stage, and replace the micromanipulator.
9. Pick up cells from the strip, and place them on the main slab, next to the spores previously set down. Use the marks on the edge of the slab, and those beside the spores, to locate the positions. The cells and spores will probably not be close to each other when they are first set down. Pick up the spore and cell together and set them down again, until they are held tight together by surface tension. Repeat until all the spores are paired with cells.
10. Transfer the slab to a plate of YEP-glucose agar as before. If possible, observe the pairs every 1-2 h to see if fusion has occurred. The slab can be lifted off the agar and placed on a sterile slide on the chamber, or if a relatively high-powered binocular dissecting microscope is available, they can be observed directly, without removing the slab from the Petri dish.
11. Incubate the cultures until colonies become visible.

c) Detection of Post-Meiotic Segregation (Fogel et al. 1983)

This procedure requires replica plating of the single-spore clones directly to selective media without disturbing the colonies or mixing the cells after germination of the spores on the spot where they were set down after dissection. Post-meiotic segregation occurs in the first division after spore germination, so that a sectored colony results. Hence the spore clones cannot be picked off a slab to a plate, and the expression of the different chromosomal markers determined, except with considerable difficulty. The least laborious method is to dissect the asci directly on the surface of the agar of the plate which will be used as the master, rather than on a slab which is then transferred to the surface of the plate. There are two methods of accomplishing this:

i) Procedure 1: Dissection on the Surface of an Open Plate

Required

1. Micromanipulator.
2. Dissecting microscope, binocular, having at least 50% and preferable 100x magnification, and at least several centimeters space between the objective and the stage.
3. Microloop instead of flat-ended needle. A microforge is necessary for making microloops, and the most satisfactory way of obtaining good ones is to have glassblower make them. However, a flat-ended needle can be used if microloops are not available.
4. Digested spore preparation, small loops, other equipment as before.
5. Plates of YEP-glucose agar (usually) for dissection.

Procedure

1. Mark the agar (needle) in a series of radial lines about 1–1.5 cm long, around the edge of the plate (Fig. 5) and mark each line with four cross cuts.

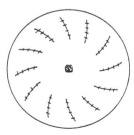

Fig. 5. Dissecting on open plates – Queen Mary College method

2. Place a loopful of spore digest in the center of the plate and spread over a small circle.
3. Place the plate on the microscope stage and focus on the spores using as high a magnification as possible.
4. Mount the microloop in the micromanipulator and adjust to where the loop can be seen in the field of the micromanipulator.
5. Locate and pick up a suitable tetrad with the loop.
6. Move the plate so that the loop and spores are over the end cross-cut of one of the lines.
7. Set down the tetrad and separate the spores, by picking them up and setting them down. Transfer the individual spores to the cross-marks.
8. Repeat until all of the marks have been filled.
9. Remove the micromanipulator and loop, remove the plate from the stage, cut out the area with the original spore digest and discard, and incubate the plate as before, until good colony growth has occurred.
10. Replica plate (velvet or filter paper) to selective media, incubate and determine the frequency of post-meiotic segregation.

Note

There are two disadvantages to this method:
1. The plate is exposed to the air, uncovered, during the whole of the dissection procedure, so that the risk of contamination is relatively high, and a clean room is essential.
2. Making microloops requires access to a microforge, an expensive piece of equipment which is not always available.

ii) Procedure 2: Dissection on an Inverted Plate (Fogel's Procedure)

Required

1. A special plate holder to hold the plate in an inverted position on the microscope stage during the dissection procedure (see also Figs. 6 and 7).
2. Flat-ended needles having the tips drawn out to about 1.5-2 cm, to accommodate the rim of the plate and reach the surface of the agar.
3. Other materials as before.

Procedure

1. Streak the spore suspension with a small loop along a diameter of the plate. The exact location of this streak depends on the type of stage, as the range of movement may not allow full access to the plate while it is held in one position only.

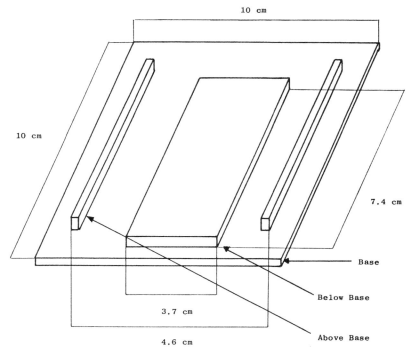

Fig. 6. Petri dish holder for dissection, Goldsmiths' College pattern

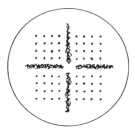

Fig. 7. Dissection pattern using Goldsmiths' College plate holder

2. Invert the plate and place in the holder on the microscope stage, with the streak running "lengthwise" in relation to the operator. Focus the low-power objective on the streak.
3. Mount the needle in the micromanipulator (it may be necessary to turn the needle through 90° to place it under the plate) and locate it in the microscope field. Bring the needle close to the agar surface and make any necessary marks as guidelines.

4. As in dissecting on a slab, pick up an ascus, move the plate with the stage controls until the needle is well clear of the streak (at least 1-2 cm).
5. Set down the ascus, dissect and set out the spores in a line perpendicular to the streak.
6. Move the stage 3 mm along the streak, marking the spot, and repeat the procedure. Continue until the available space has been used, then repeat on the other side of the streak, assuming that the stage has sufficient range of movement. It should be possible to dissect at least 40 asci on one plate. If the range of movement of the stage is somewhat restricted, divide the plate into four sectors, make the streaks along one of the radii, and dissect to one side only.
7. Remove the plate from the holder and cut out the original streak(s) and discard. Make a "cookie cutter" for this, from a strip of thin sheet metal, if desired.
8. Incubate the plate (25-30 °C) for 2-3 days until the colonies are a suitable size, and replica plate to selective media for detection of postmeiotic segregation.

Note

A rather sophisticated plate holder, micrometer stage controls and stage-mounted needle holder are now commercially available. A simplified version can be built in any reasonably good workshop.

REFERENCES

Fogel S, Mortimer RK, Lusnak K (1983) Meiotic gene conversion in yeast: molecular and experimental perspectives. In: Spencer JFT, Spencer Dorothy M, Smith ARW (eds) Yeast genetics: Fundamental and applied aspects. Springer, Berlin Heidelberg New York, pp 65-107

Sherman F (1973) Appl Microbiol 26:829

4. Enzymes for Digestion of Yeast Ascus Walls

Snail enzyme (glusulase, helicase), zymolyase, Novozyme, etc., are all satisfactory. These enzymes can also be used for removal of walls from vegetative yeast cells (see section on protoplasts and protoplast fusion, and on transformation). A suitable enzyme for digestion of ascus walls can be extracted from commercial mushrooms, as follows:

1. Cut up 450-500 g of commercial mushrooms and homogenize in 500 ml of distilled water.

2. Strain off solid residues through gauze.

3. Centrifuge the liquid at 10-20,000 g and collect the supernatant and allow to stand at 4 °C overnight.

4. Centrifuge again at high speed.

5. Collect supernatant and sterilize by filtration. Cellulose membrane filters are not suitable, for obvious reasons. Use asbestos filter mats or similar material (Ford Sterimat or Seitz filters).

6. Store frozen, in 10-20-ml aliquots.

7. The crude enzyme can be purified by standard methods for protein purification.

5. Ancillary Methods for Handling Clones Obtained by Ascus Dissection

a) Replica Plating

Methods for this technique are well-known. A base, holding the velvet pad, is preferable to a hand-held "stamp", which often leads to very smeary plates. Either velvets or filter paper can be used, the latter giving more sharply defined colonies (Fig. 8).

i) Velvet Pad

Required

1. Velvet squares, approximately 20 × 20 cm, wrapped in foil and sterilized.
2. Agar plates of selective media.

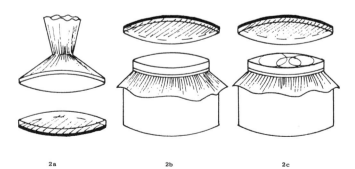

Fig. 8. Replica plating

3. Base, with circular top, to fit inside of Petri dish when velvet is fitted over it. The velvet is held in place with a heavy rubber band or a plastic or metal ring.
4. Master plate (from mutagenic treatment or picked from dissections) having 100-200 colonies/plate or up to 32 streaks.

Procedure

1. Mark the base of all plates with an index mark for alignment.
2. Partly open the pack of sterile velvets. Take out one, by one corner, and lay it over the top of the mushroom, pile side up. Hold in place with rubber band or ring. Do not touch surface.
3. Open the master plate, hold index mark in a known position, and press down on the velvet, making sure of even contact everywhere.
4. Remove the master plate, cover and set aside.
5. Repeat procedure with each of the plates of selective media.
6. Label all plates, incubate, and read if expression of auxotrophic requirements is desired (clones from dissections). If mutants are to be selected, compare selective plates with master and select those not growing on the elective media.

ii) Filter Paper

Required

As above, but also 7- and 12-cm filter papers in a fresh pack, preferably. Whatman No. 1 or equivalent quality is satisfactory. Tweezers, alcohol for flaming.

Procedure

1. Clamp a velvet on the base as before.
2. Cover the velvet with three circles of 7 cm diameter filter paper, to absorb excess moisture.
3. Cover the 7-cm filter papers with a 12-cm diameter filter paper (use aseptic techniques, and if possible, a fresh box of filter paper or one which has been kept closed and reserved for the purpose. Generally, filter papers which have not been handled or exposed to room air can be considered sterile, for this purpose). Clamp to the base as before.
4. Invert master plate on the filter paper and press down. Follow by plates of selective medium as before.
5. Discard papers, and repeat with next set of master and selective plates. Incubate as before.

Note

Replica plating by the filter paper method is said to cause less smearing of the colonies than when velvets are used. Sterilization of the filter papers may be done in a large petri dish or other container if necessary.

b) Mating Type Determination by Cross-Stamping

For determination of the mating type of only a few unknown strains, or of prototrophic haploids, only two mating-type tester strains of known mating types, and plates of YEP-glucose agar, are required. Fresh cultures of the strains are mixed on a plate, and incubated at 30 °C. After 3–6 h, water mounts of the mixtures can be observed microscopically and the characteristic shapes of zygotes, where mating has taken place, can easily be recognized. If desired, the zygotes can be isolated by micromanipulation and pure diploid strains grown up from them.

Determination of mating types of auxotrophic clones, isolated from dissected asci, is quite laborious if done by the above method. Determination of the mating types of large numbers of such clones is best done be crossstamping with haploid auxotrophic strains of known mating type as follows:

Required

1. Plates of YEP-glucose agar and YNB-glucose agar.
2. Sterile strips of wood, cut to length to extend past eight mini-streaks on a master plate. Wooden tongue depressors, if available, can be cut to the correct length, and can be repeatedly autoclaved.
3. Known haploid mating types, auxotrophic for complementing requirements to the unknown strains. A set of all of the adenine auxotrophs is most useful and highly desirable as tester strains.
4. Velvets or filter paper, and all other materials required for replica plating.
5. Sterile toothpicks, tweezers, alcohol and similar ancillary materials as required.

Procedure

1. The mating tester strains are previously streaked on a plate of YEP-glucose medium, the streaks being slightly longer than the wooden strips, and grown for 24–48 h.
2. The strains to be tested are also placed on YEP-glucose plates in sets of four (assuming that these strains are clones from dissected asci). Up to 32 strains can be streaked on a plate (see Fig. 9).

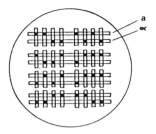

Fig. 9. Cross-stamping pattern for determination of mating type

3. At the same time, take out one of the wooden strips, using aseptic precautions at all times, dip it in the streak of tester strain, touch it gently a few times to the surface of the agar to remove excess cells, and then press it firmly across the first set of eight mini-streaks on the other plate, using the upper part of the small streaks. Wipe the edge of the wooden strip with a tissue and discard it to another jar so that it can be re-autoclaved later.
4. Take another wooden strip and repeat the process until all four rows of mini-streaks have been cross-stamped on the upper part.
5. Change to the other mating type of the same tester strain, and repeat again until all four sets of mini-streaks have again been stamped, this time on the lower half. The result is a set of unknowns, in four sets of eight, which have all been stamped at right-angles with two streaks per row, of the two mating types of each tester.
6. Incubate this master plate for 24 h or until reasonable growth is obtained.
7. Replica plate the master to suitable dropout media to permit growth of any diploids which have resulted from possible matings where the streaks cross. YNB-glucose is the most suitable minimal medium, if the auxotrophic requirements of the tester strains have been matched to the probable requirements of the unknowns correctly. After a further 24–48 h of growth, patches of prototrophic growth should occur in zones where mating has taken place (Fig. 9).

Notes

1. The cross-stamping method can be used for unknown prototrophic strains as well, though unless there is some difference in growth rate, or, for instance one of the strains is a petite mutant carrying a cryptic mutation for antibiotic resistance, and the other is a grande form which is sensitive to the antibiotic, and the master plate is replica plated to YEP-glycerol agar containing the antibiotic, the only way in which mating can be detected is by microscopic examination of the cells at the junction between the tester and the unknown strain.

2. A complete set of adenine auxotrophic tester strains could be obtained from the Department of Genetics, University of Washington, Seattle, Washington, USA.

6. Mutagenesis

a) Ultraviolet and X-Irradiation

These are probably the simplest means for induction of mutations in yeast, since there is no problem of disposal of toxic wastes. In particular, UV light is usually readily available, and does not have the same penetrating power as X-rays. Recessive mutations are usually induced in haploid strains, as they are not normally expressed in diploids, but dominant mutations (usually) to resistance to some toxic compound can be induced in strains of any ploidy. Recessive mutations can be induced in aneuploid (industrial) yeast strains, but are difficult to maintain. The strains are apparently unstable and revert to prototrophy readily.

Mutagenesis by UV light can be carried out either in liquid or on solid medium.

i) Procedure 1: (Liquid Medium)

Required

1. Cultures for mutagenesis. Usually haploid, for reasons given above. Strains are grown in liquid YEP-glucose medium for 24-48 h.
2. YEP-glucose and YNB-glucose (minimal) agars.
3. Sterile water blanks, 9.0 and 9.9 ml.
4. Sterile Petri dishes, empty.
5. Germicidal lamps, in a case to keep UV light from reaching the eyes. If there is any need to work in UV light, safety glasses, UV-proof, are required.
6. Small rotary shaker, if available.
7. Timer. An irradiation chamber with a shutter and automatic timer is best, but the exposure may be controlled with a card which can be removed and replaced by hand.
8. Ultrasonic apparatus, for breaking up clumps of cells.

Mutagenesis

Procedure

1. Dilute the culture 1:10, into a water blank. Switch on UV lamp, to warm up.
2. Sonicate diluted culture lightly (10-15 s at low power) to break up clumps and separate any buds from the mother cells.
3. Pour the suspension into a Petri dish and place on the shaker, if one is available, in the irradiation chamber. If there is not an automatic shutter, cover the dish with a card.
4. Set timer, start shaker (low speed, to keep cells in suspension) and begin exposure. Either set shutter to open or remove card. Use exposure length of 15 s to 15 min, depending on the strain and species of yeast.
5. Agitate Petri dish by hand if a shaker is not available. Dose may be 18 ergs/min/mm. Close shutter or replace card.
6. Remove Petri dish from irradiation chamber.
7. Either (a) remove cells and resuspend in YEP glucose broth and incubate 2-3 h to permit a round of cell division before plating or (b) dilute irradiated culture directly into water blanks at appropriate dilutions ($1:10^3$, $1:10^5$ and $1:10^6$).
8. Plate desired dilutions on YEP-glucose agar.
9. Incubate for 2-5 days at (normally) 30 °C.
10. Replica plate colonies to minimal medium, if auxotrophic mutants are desired, or to other selective media. Incubate another 2-5 days, or longer if necessary, retaining the master plates.
11. For auxotrophic mutants, examine plates to find those colonies which have grown on the master plates but not on the minimal medium. Pick selected colonies to complete medium and save. Test on dropout media to determine the exact nature of the mutations.
12. For resistance mutants, examine plates to find colonies which have grown.

ii) Procedure 2: (Solid Medium)

Required

1. Cultures for irradiation, as before.
2. YEP-glucose and YNB-glucose (minimal) agar, as before.
3. Sterile waterblanks.
4. Irradiation chamber.
5. Ultrasonic apparatus.

Procedure

1. Switch on UV lamp to warm up.
2. Dilute culture 1:10, as before.
3. Sonicate lightly to break up clumps and separate buds from mother cells.
4. Make appropriate dilutions and spread aliquots on YEP-glucose plates. Allow plates to dry.
5. Place plates in irradiation chamber and remove cover. Start timer.
6. After irradiation is finished, replace cover, switch off lamp and remove plates.
7. Incubate plates 2–5 days at 30 °C.
8. When colonies have all appeared, replica plate to minimal medium or other selective medium as before, and select those colonies which do not grow on the plates.
9. Restreak and purify colonies, and inoculate on to dropout plates to determine the requirements of the auxotrophs.

Notes

1. For isolation of mutants resistant to toxic materials, replicate master plate to medium containing the compound in question.
2. Particularly for procedure (2), it is desirable to spread several plates of each dilution to allow several irradiation times to be used. The degree of killing should be approximately 90%.
3. Some of the more sophisticated irradiation chambers may contain a UV lamp (germicidal), a shutter and timer for automatic exposure control, with space for a shaker and plate holder, to keep the yeast suspension well agitated. For less fortunate investigators, a shielded box to hold the UV tube is satisfactory, and the plate of suspension can be agitated by hand.

iii) General Note on Mutagenesis

In the search for mutations, the purpose for which the mutants is required should always be kept in mind. If a particular type of mutant is required, which has not been reported before, or if the effect of a particular mutagen is under investigation, then it may be necessary to proceed with mutagenic treatment of a suitable yeast strain, and isolate and characterize the mutants thus obtained. However, if it is desired to investigate some genetic event such as gene conversion, recombination mechanisms, cell division cycle phenomena, and many other similar subjects, it may be cheaper and easier to acquire a suitable mutant from a colleague or from a reputable collection. This is especially true where the effect of some particular mutation is under investigation, or where a strain carrying several mutations at known sites is required. The collection at the Genetic Stock Center, Department of Medical Physics, University of California, Berkeley, CA94720, USA, is a particularly good one, and strains are available at a reasonable cost.

X-Irradiation. Follow procedures as for UV-irradiation, on solid or in liquid media, as desired. For the actual irradiation, follow the instructions for the machine available, avoiding irradiation of the operator and/or onlookers, wearing any protective clothing specified, and, if possible, getting advice and assistance from an experienced operator. The machine should be checked regularly for radiation leakage and general safety.

iv) Special Note: Induction of Mitotic Recombination by Ultraviolet Irradiation

Low doses of irradiation (1000-2000 ergs/cm^2) will induce mitotic recombination in diploid yeast strains carrying recessive mutations. These mutations can then be expressed and determined by standard methods (plating on selective media). The requirements and procedures are the same as for induction of auxotrophic and resistance mutations in haploid yeasts, except that the UV dosage is much lower (killing rate only a few percent). When the colonies have grown up after irradiation of the original cells, they should first be examined visually for morphological evidence of sectoring, after which the colonies are replica plated to minimal media and auxotrophic sectors are selected and purified for characterization. The method has been used for detection of naturally-occurring recessives to auxotrophy in *Candida albicans*, (Whelan et al. 1980; Sarachek et al. 1981) as well as for *Saccharomyces cerevisiae*, demonstrating incidentally the diploid nature of the former species.

b) Drying

Studies of the effect of drying are incomplete, but indicate that in combination with the presence of oxygen, it may induce specifically, mutations to adenine auxotrophy, particularly at the ade1 and ade2 loci. Both freeze-drying and vacuum-drying induced mutations at these loci, the mutation rate being greater for loci more distant from the centromere. Thus, the genetic change is probably due to crossing-over between the locus and the centromere. Gene mutation and gene conversion were not observed.

i) Freeze-Drying

Required

1. Washed cell cultures, age 18-20 h, grown in complete medium.
2. Suspension medium (skim milk, 5% + 0.5% sodium glutamate).
3. Small test tubes (11 mm diameter).
4. Freezing mixture (dry ice in acetone).
5. Vacuum pump and cold trap, manifold for connecting tubes to pump.
6. Plates of synthetic complete medium or YEP-glucose agar, minimal agar (YNB-glucose), and appropriate dropout media for characterizing auxotrophic mutants.
7. Water blanks, 9.0 and 9.9 ml, sterile, for making dilutions and plating cells to determine total survivors and numbers of mutants.

Procedure

1. Pipette 0.1 ml of cell suspension (3×10^6 cells/ml) into the test tubes.
2. Freeze in dry ice-acetone for 10 min ($-20\,°C$).
3. Connect the test tubes to the vacuum pump (through the trap!) and evacuate. Dry for 2 h.
4. Resuspend cells in complete medium, dilute appropriately and plate on complete medium agar.
5. Incubate for 2-5 days.
6. For adenine mutations, count red and sectored red colonies, and isolate any desired number.
7. For other mutations, replica plate to minimal medium, isolate auxotrophic strains, and characterize on dropout media.

ii) Vacuum Drying

Required

1. Membrane filters, 47 mm diameter (millipore type).
2. Vacuum system as for freeze-drying.
3. Washed yeast cultures, as for freeze-drying.
4. Synthetic complete or YEP-glucose agar, minimal medium (YNB-glucose agar), and dropout media as in procedure for freeze-drying.
5. Water blanks, 9.0 and 9.9 ml, sterile, for making dilutions for plating cells for determination of total survivors and mutants.

Procedure

1. Filter suspension and collect cells on the membrane filter.
2. Dry the filter and cells for 2.5 h in the dark or in dim yellow light.

3. Resuspend cells in synthetic complete medium, dilute appropriately, plate on CM agar and determine total numbers of surviving cells and mutant clones as before.

Note

Menadione (vitamin K_3) enhances the toxic effect of oxygen on yeast (Chaput et al. 1983), though its action on mutagenesis in yeasts by oxygen under conditions of drying has not been determined. The compound may, however, prove to be useful in enhancing mutagenesis as well.

REFERENCES

Chaput M, Bruggier J, Lin Y, Sels A (1983) Biochimie 65:501–512
Sarachek A, Rhoads DH, Schwartzhoff RH (1981) Arch Microbiol 129:1–8
Whelan WL, Partridge RM, Magee PT (1980) Molec Gen Genet 180:107–113

c) Chemical Mutagens

The general method is to suspend the cells in a solution of the mutagen and incubate for long enough to induce the maximum yield of mutations, after which the mutagen is preferably neutralized, the cells are recovered, and the mutant clones are selected. The cells may, as with UV and X-irradiation, be incubated in complete medium for a round of cell division to fix the mutations, before they are plated and selection of the mutants is begun.

Mutagens known to cause specific classes of mutations can be grouped as follows:

1. Base analogues. 5-Bromouracil (5-BU), 5-bromo-deoxyuridine (5-BUdR). These compounds cause replication errors when incorporated into the DNA, in the rare enol form, so that G is incorporated instead of A. 5-BU can induce reversions of mutations of this type. 2-Aminopurine has similar effects.

2. Deaminating agents. Nitrous acid (NA) deaminates A to hypoxanthine, and hydroxylamine converts C to a derivative that pairs with adenine, producing a GC to AT transition, not reversible by NA.

3. Alkylating agents. EES, EMS, MMS, nitrosoguanidine, ICR-170, and numerous others, also produce transitions.

3a. Agents producing transversions (point mutations), which are not reverted by agents inducing transitions or frameshifts. These mutagens include 4-nitroquinoline-1-oxide, which also induces other types of mutations. Mn^{2+} also induces transitions.

4. Acridine derivatives; ethidium bromide. These cause frameshift mutations, which can be reversed by acridine, or by a second frameshift, but not

by mutagens of the previous groups. Acridines and ethidium bromide, and Mn^{2+}, preferentially cause mutations in mitochondrial DNA in yeast.

5. Cross-linking agents, including nitrous acid, bifunctional alkylating agents, and irradiation, cause deletions of varying lengths, in the DNA.

6. "Spontaneous" mutations, which arise primarily from enzymatic imperfections during DNA replication or recombination and repair, consist primarily of frameshifts, deletions and transversion.

Procedures for treatment with MNNG (N-methyl-N'-nitro-N-nitrosoguanidine), EMS (ethyl methane sulfonate), MMS (methyl methane sulfonate), NA (nitrous acid) are given as examples.

i) Nuclear Mutations

Required

1. Yeast cultures, grown to stationary phase in YEP-glucose broth. Use haploid strains if recessive mutations (auxotrophs) are desired. Diploid strains may be used for investigation of dominant mutations, if desired.
2. Phosphate buffer, 0.05 M, pH 7.1.
3. Mutagens (final concentrations, MNNG, 25 µg/ml; EMS, 3%; MMS, 0.04% (in 0.1 M buffer); NA, 2 mg/ml, in 0.1 N sodium citrate buffer, pH 4.5.
4. $Na_2S_2O_3$, 5%, for terminating the reactions with MNNG, EMS or MMS.
5. Complete and selective media as before.

Procedure

1. Wash the yeast cells in sterile buffer.
2. Resuspend the cells in buffer containing the desired mutagen at the specified concentration.
3. Incubate 10–40 min, depending on the mutagen, at 25–30 °C.
4. Dilute the suspension into 5% $Na_2S_2O_3$ to terminate the reaction (MNNG, EMS, MMS) or into 0.1 M phosphate buffer for NA.
5. Make appropriate dilutions of the cell suspension, and plate aliquots on complete and selective media. Incubate for up to 5 days. Select auxotrophic mutants by replica plating to minimal medium and selecting colonies which fail to grow.

ii) Mitochondrial Mutations

1. Petites. The frequency of spontaneous petite mutants in *Saccharomyces cerevisiae* is relatively high, reaching 100% in some strains. These can be selected by colony size and tested by streaking on media containing glycerol or lactate as sole carbon source (YEP-glycerol or YNB-glycerol, for instance).

a) Petite mutations induced by starvation. The frequency of spontaneous petites can be induced in some strains by incubating them for one or 2 weeks in a very weak medium (Schöpfer's medium, see Notes), followed by plating out on YEP-glucose agar or some similar medium.

b) Heat. In some strains, growth or holding at elevated temperatures increases the frequency of petite mutants.

c) Mutagenesis by chemical agents (acriflavin, ethidium bromide, Mn^{2+} ions and some other agents. Acriflavin and ethidium bromide are the mutagens of choice. Both are intercalating agents, and both affect mitochondrial DNA preferentially. Acriflavin acts on mtDNA in growing cells, and ethidium bromide on both growing and non-growing cells. Ethidium bromide is more drastic in its effects, and may bring about complete destruction of all mtDNA in the cells. Petites made with acriflavin, on the other hand, may contain enough non-functional mtDNA to be plainly visible in cells stained with DAPI (4',6-diamino 2-phenyl-indole), and this DNA may recombine with incoming mtDNA from other yeasts with which the petite strain is hybridized. Mn^{2+} is more often used to induce mutations to antibiotic resistance in yeast, but produces a high frequency of petite mutants as well. Induction of petites by chemical mutagens apparently causes a greater disruption of the mtDNA than is found in spontaneous petites, which usually contain mtDNA of the normal size, but which is formed by reduplication of a small segment of the original mtDNA, so that most of the normal mitochondrial genes are lost. However, mutations to antibiotic resistance, encoded on mtDNA, may be retained on the reduplicated segment, and are again expressed when the petite is mated with a sensitive grande strain.

Required

Acriflavin

1. Tubes of YEP-glucose broth, 5 ml/tube.
2. Acriflavin solution, 0.4%.
3. Standard loops.
4. Plates of YEP-glucose and (later) YEP-glycerol agar.
5. Slants of the desired yeast cultures.

Procedure

1. Add a loopful of the acriflavin solution to each of the YEP-glucose tubes required.
2. Inoculate the tubes from slopes of the yeast cultures.
3. Incubate at 30 °C for 2-3 days, without agitation, in the dark.

4. Streak the cultures on YEP-glucose agar and incubate for 3-4 days, or until the colonies are large enough so that the size differences can be seen.
5. Pick small colonies to YEP-glucose and YEP-glycerol agar (toothpicks). Petites fail to grow on YEP-glycerol medium, and can be identified on the YEP-glucose plate and stored.

Antibiotic-Resistant Mutants. Mutants of this type, which are frequently of mitochondrial origin, may be obtained by spontaneous mutation or by treatment with chemical mutagens. Spontaneous antibiotic-resistant mutants may be isolated by spreading a heavy suspension of cells of the desired strain, either recovered from a broth culture or scraped off the surface of a plate or slant, on plates of YEP-glycerol medium containing the antibiotic. Concentrations of the more commonly used antibiotics are as follows: chloramphenicol and erythromycin, 1.5-2.5 mg/ml, oligomycin 2.5 µg/ml. Numerous other drugs may be used for special purposes (daunomycin, chlorimipramine, trimethoprim, adriamycin, and many others. Some of these must be handled with the precautions required for operations with carcinogenic agents).

Mutagenesis with Manganese Ions (Putrament et al. 1977)

Required

1. YEP-glucose broth; 5 ml in tubes.
2. $MnCl_2$ solution, 79 mg/ml.
3. YEP-glycerol plates, containing chloramphenicol, erythromycin, oligomycin, or other antibiotics (required when growth appears in the tubes).

Procedure

1. Add 0.1 ml of the $MnCl_2$ solution to each of the tubes of YEP broth. Higher or lower concentrations may be necessary, depending on the individual yeast strain and its tolerance of Mn ions.
2. Inoculate the tube with the desired yeast strains, using a relatively small inoculum.
3. Incubate on a shaker or roller drum for 7-10 days, or until growth is moderately heavy. Growth in the presence of Mn ions is often slow and clumpy.
4. Spread approximately 0.5 ml of the culture on the antibiotic plates. Washing of the yeast cells is not necessary. Allow the plates to dry at room temperature and incubate at 25-30 °C until colonies appear. This may take up to 2 or 3 weeks, again depending on the strain and its response to the Mn treatment.

5. Restreak the colonies on the appropriate antibiotic-containing medium and pick to slants. Usually three or four colonies per plate are sufficient; in theory, one per plate is all that should be isolated, but in practice, we have found that the strains thus obtained do differ in their characteristics.

Notes

Schöpfer's medium, contains, per liter, glucose 30 g, asparagine 1 g, KH_2PO_4 1.5 g, and $MgSO_4 \cdot 7 H_2O$, 0.5 g. Five-ml batches, in culture tubes or McCartney bottles, are inoculated with the desired strains and incubated at 30 °C for 6-10 days. The cells are recovered and plated on YEP-glucose, and small colonies are picked to YEP-glucose and YEP-glycerol, and any petites formed are selected.

REFERENCE

Putrament A, Baranowska H, Prazmo W (1973) Molec Gen Genet 126:357-366

d) Isolation of Particular Mutants or Classes of Mutants

i) Mutator Mutants

These are mutants which show an increased rate of spontaneous mutation (6-20 times normal), at various loci. One of the more sophisticated methods for isolation and study of these mutants is that used by R.C. von Borstel et al. 1971):
1. Detection of mutator mutants, using lysine auxotrophs.
The yeast strains are grown in normal medium and washed.
10^6-10^7 cells are spread on one plate of MC (synthetic complete medium containing 20 μg/ml of lysine), and on lysine omission medium.

Incubate for several days (5-7) until the yeast has grown in a thin lawn, and has used up all the lysine, and only the revertants to lysine independence can continue to grow. Strains containing mutator mutants show a much higher frequency of colonies on MC plates.
2. Determine revertant frequency.

Required

1. Limiting medium (e.g. lysine, 1 μg/ml, uracil, 0.5 μg/ml), made up with all other requirements at normal levels.
2. Yeast strain, grown as before in normal medium.
3. Plastic culture boxes, (10 × 10 wells, presterilized, 10-12 boxes per strain to be analyzed.

4. Automatic dispensing apparatus (Brewer automatic pipettor or other similar type), set to deliver 1 ml of medium per stroke.
5. Flask, to hold entire batch of inoculated medium with stirring bar, sterile.
6. Magnetic stirrer.

Procedure

1. Inoculate the flask of limiting medium (approximately 1.5 l with 5 × 10^3 cells/ml. Start magnetic stirrer and let system equilibrate.
2. Connect flask to automatic pipettor (aseptic!).
3. Start automatic pipettor, deliver 20 aliquots into a 25-ml graduated cylinder to check calibration.
4. Immediately begin filling wells, 1 ml in each, so that 10–12 boxes are filled for each yeast strain to be tested.
5. Plate aliquots on solid medium to determine numbers of revertants in the original inoculum.
6. Deliver another 20 aliquots to a graduated cylinder for a further check on the volume delivered.
7. Seal each box with masking tape and incubate at room temperature (24–25 °C). Handle boxes gently to avoid any agitation or spillage between wells.
8. After incubation, count wells having no colonies and record numbers. Count also total number of reverted colonies.
9. Determine total numbers of cells by counting the cells in two wells/box, in a hemocytometer.
10. Revertant colonies (one per well) are removed and plated on complete, synthetic complete, minimal and omission medium (without the other auxotrophic requirements) to determine whether the reversion were at the *lys*1 (e.g.) locus (growth on lysine omission medium only) or were due to a super-suppressor.
11. The mutation rate is calculated as follows: If N is the total number of compartments in an experiment, and N_0 is the number of compartments without revertants, m is the average number of mutational events/compartment, m_b is the average number of mutants/compartment, as determined by direct plating, in the original inoculum, m_g is the corrected number of mutational events, then

$$e^{-m} = N_0/N$$

and $m_g = m - m_b$, so that the mutational events/cell/generation, $M_1 = m_g/2C$, where C is the number of cells/compartment after growth has ceased in the limiting medium (the number of cell generations in the history of a culture is approximately twice the final number of cells).

12. If the mutants are to be grouped into categories, the mutation rate is $M_i = f_i M$, where M_i is the mutation rate (per cell per generation) and f_i is the fraction of mutants tested which were found in the ith category.

Notes

1. The cultures originally used were radiation-sensitive mutants derived from a multiple auxotroph requiring adenine (ade2-1), arginine (arg4-17), histidine (his5-2), lysine (lys1-1), and leucine (leu1-12). The strains were not given further mutagenic treatment; i.e. the value determined was the spontaneous mutation rate. All of the radiation-sensitive single-spore clones proved to have mutator capacity.
2. Other mutator strains can be induced by treatment with EMS, according to procedures given earlier. Many of these had little or no sensitivity to radiation.

Procedure for Obtaining Mutator Genes by Mutagenic Treatment

1. The strain used had the genotype *trp*5-48 *his*5-2 *arg*4-17 *lys*1-1 *ade*2-1, all of which were super-suppressible mutants (ochre), and gives red colonies on complete medium.
2. After treatment with the desired mutagen, according to the procedures previously described, the mutagen was neutralized in thiosulfate and the cells were washed in buffer (1/15 M KH_2PO_4), diluted into 10 ml of YEP-glucose broth and 1 ml of this was placed in eight tubes containing 10 ml of YEP-glucose broth, shaken and incubated overnight.
3. These suspensions were diluted appropriately and plated on YEP-glycerol agar to eliminate petites, and incubated for 15 days to allow papillae to form.
4. Colonies with more than seven papillae were scooped up (sterile applicator) and suspended in 2 ml of 1/15 M KH_2PO_4 solution.
5. An 0.5-ml aliquot was plated on MC (synthetic complete) medium, and the number of mutant revertants was counted. Isolates showing a doubling of revertants over the control were collected and restreaked (cell counts were not necessary, as the final background count is the same in all cases).
6. The isolates were restreaked on YEP-glucose agar and one colony was taken, suspended in water or buffer, and aliquots were plated on minus-lysine, MC and YEP-glucose agar, and medium containing p-fluorophenylalanine (MC with 3x amino acids, and 250 µg/ml of p-fluorophenylalanine).
7. Strains still showing double the numbers of lysine revertants and/or twice as many p-fluorophenylalanine mutants were presumed to have a mutator gene and were used for determination of the mutation rate as previously described.

Note

Both cytoplasmic and nuclear petite strains may exhibit the enhancement of the spontaneous mutation rate.

REFERENCE

Von Borstel RC, Cain KT, Steinberg CM (1971) Genetics 69:17−27

ii) Temperature-Sensitive Mutants

These mutants are relatively easy to obtain, and have been widely used in studies of the cell division cycle. After mutagen treatment, the cell suspension is diluted and spread on complete medium (YEP-glucose) and incubated. After incubation, plates having approximately 50−100 colonies/plate are replica plated to two YEP-glucose plates, one of which is incubated at the restrictive temperature (usually 34−37 °C) and observed at intervals. Cells may be fixed and stained (Giemsa) before and after transfer to the restrictive temperature, to determine the terminal phenotype, and time lapse photomicrography can be used to classify the mutation according to the number of cycles the cell goes through before arrest. Colonies failing to grow, after one or more cycles of cell division, at the restrictive temperature can then be further characterized according to the point in the cycle at which development is halted.

Cold-sensitive mutants may be isolated in a similar way.

Osmotic-remedial mutants may be detected and isolated by replica plating *ts* strains to YEP-glucose agar containing 0.5, 1.0 and 1.5 M KCl, and observing those which now grow at the restrictive temperature.

iii) Isolation of *kar* Mutants (Defective in Nuclear Fusion)
(Conde and Fink 1976)

These mutants, in which nuclear fusion after karyogamy does not take place, either in normal mating or during protoplast fusion, are extremely useful in transferring cytoplasmic elements (cytoduction) with minimal interference from nuclear genes, and also in transfer of single nuclear chromosomes. The simplest way of obtaining strains carrying this mutation is to buy or borrow them, but failing that, the following procedure may be used.

Required

1. Yeast strains, carrying several easily observed nuclear mutations, including mutations to canavanine and/or cycloheximide resistance. Strains JC1

(α his4 ade2 can1 nys^R ρ^-) and GF4836-8C (a leu1 thr1 ρ^+) have been used in the construction of *kar* mutants.
2. EMS as mutagen.
3. Media: YEP-glucose broth and agar.
   YNB-glucose media (minimal, without amino acids; MM).
   Selective medium. MM + histidine (0.3 mM), adenine (0.15 mM), glycerol 3%, glucose 0.1%, buffered with 0.1 M citrate-phosphate buffer (pH 6.5) added (sterile solution, 10x strength), after sterilization. Canavanine (60 mg/l) and nystatin (2 mg/l) are added when the autoclaved medium had cooled to approximately 50 °C.

Procedure

1. Treat strain JC1 with EMS according to methods given previously.
2. Cross the two strains by mass mating (10^8 cells/ml of JC1 with 10^9 cells/ml of GF4836-8C).
3. Spread mating mixture (0.2 ml) on YEP-glucose plates and allow to mate for 20-50 h at 30 °C.
4. Resuspend cells from each plate in 1 ml of water, dilute 1:10 and spread 0.2 ml/plate on selective medium.
5. Isolate colonies appearing on the plates and test. These colonies will probably be mostly α his$^-$ ade$^-$ can^R nys^R (rho$^+$) (JC1 nucleus and GF4836 mitochondria) since the two parent strains cannot grow on this medium, being auxotrophic for two markers, and one is petite and cannot grow on glycerol as sole carbon source, and the other is sensitive to canavanine and nystatin.
6. Test the strains thus obtained by back-crossing to GF4836-4C (not mutagenized) and determining the frequency of heteroplasmon formation (use the cross, JC1 × GF4836-4C, as a control).

REFERENCE

Conde J, Fink GR (1976) Proc Natl Acad Sci USA 73:3651-3655

iv) Cell Wall Mutants

Mutants Forming Mannans with Altered Chemical Structures. These mutants have been used by Ballou et al. (1980) and Cohen et al. (1980) to investigate the biosynthesis and chemical structure of the cell wall mannans of *Saccharomyces cerevisiae* and *Kluyveromyces lactis*, among other species.

Required

1. EMS, for mutagenesis. See previous section for other requirements for use of this mutagen (p. 35, 36).
2. Antisera, directed against pentasaccharide (or other mannan side-chains) side-chains of the yeast mannan, having the N-acetylglucosamine terminal group *(K. lactis)*. These antisera were obtained by adsorption of anti-*K. lactis* serum with cells of *S. cerevisiae*. Anti-*K. lactis* and anti-*S. cerevisiae* sera may be raised in rabbits, by standard immunological methods.
3. Complete medium (broth and agar plates).
4. Sterile 0.9 M NaCl solution.
5. Thiosulfate solution, for neutralization of EMS after mutagenesis.

Procedure

1. Grow the yeast for 24-48 h in YEP-glucose broth and recover cells, then mutagenize with EMS as previously described.
2. Recover the cells, wash and resuspend in complete medium and incubate (with shaking) for 2 days at 30 °C.
3. Recover and wash the cells and resuspend in sterile 0.9% NaCl solution.
4. Aggutinate the wild-type cells with the antiserum (0.5 ml) against the N-acetylglucosamine-containing pentasaccharide, for 1 h.
5. Shake the suspension and again allow the agglutinated cells to settle.
6. Transfer 0.2 ml of the supernatant to 2 ml of fresh medium and grow for 48 h at 30 °C.
7. Repeat the agglutination and growth procedures twice, to enrich the culture in mutant cells.
8. Plate on complete agar medium.
9. Select single colonies and streak on complete medium.
10. After another 24 h, test the colonies for the ability to agglutinate with the antiserum against the *K. lactis* pentasaccharide.
11. Select strains which do not agglutinate with the antiserum, and use for determination of the nature of the cell wall mannans (see Ballou et al. 1980; Gorin and Spencer 1970).
12. Mutants of *S. cerevisiae* (mnn 1-5), can also be isolated by enrichment in a similar way, using the appropriate antisera, raised in rabbits, against ethanol-killed wild-type cells, and the various mnn mutants.

Isolation of Mannoprotein Mutants by Fluorescence-Activated Cell Sorting

Required

1. Mutagenized yeast cells *(K. lactis)* of the desired strain, grown in 5 ml of YEP-glucose medium, washed and suspended in 0.9% NaCl solution

(0.1 ml) containing 0.25 mg/ml of fluorescein-labelled wheat germ agglutinin (fl-WGA).
2. Incubate for 1 h at 25 °C.
3. Recover cells, wash twice, and resuspend in saline. For screening many clones, the cells may be treated in a microtitre plate. Wild-type *K. lactis* cells bind three times as much fl-WGA as the mutants.
4. The labelled cells were separated in a modified Becton-Dickinson FACSII instrument equipped with a Spectra-Physics 171 laser (488 nm, 1 W).
5. Cells showing less than 10% of the maximum fluorescence intensity are sorted directly on to YEP-glucose agar plates.
6. The plates are incubated at 30 °C for 48 h and the resulting clones are screened for fl-WGA binding and agglutinated by specific antisera.

REFERENCES

Ballou L, Cohen RE, Ballou CE (1980) J Biol Chem 255:5986–5991
Cohen RE, Ballou L, Ballou CE (1980) J Biol Chem 255:7700–7707
Gorin PAJ, Spencer JFT (1970) Adv Appl Microbiol 13:25–89

"Fragile" Mutants. *According to Venkov* (P.V. Venkov et al. 1974). These are mutants which lyse at normal osmotic tensions, and which can be used for obtaining cells components by means of less drastic treatments than those required for breaking the walls of wild-type cells (use of lytic enzymes or ballistic cell disruption). The mutants have other characteristics which may be of interest as well.

Required

1. Materials for mutagenesis of the desired yeast strains (EMS, $Na_2S_2O_3$, etc).
2. Osmotically-stabilized medium (YEP-glucose + 10% sorbitol, or equivalent amounts of other osmotic stabilizer) (sucrose, PEG, KCl).
3. Normal YEP-glucose (2%) agar media.
4. YEP-glucose broth for production of cells.

Procedure

1. Mutagenize the yeast cells (48-h culture) according to previous procedures with EMS; stop the reaction with thiosulfate as before.
2. Dilute cells appropriately; plate on osmotically-stabilized YEP-glucose (with 10% sorbitol or other stabilizer).
3. Incubate at 30 °C for 2–3 days or until colonies have developed sufficiently. Fragile mutants may take a little longer to grow.

4. Replica plate to normal YEP-glucose agar.
5. Select colonies which fail to grow on normal YEP-glucose medium. Test for fragility and stability.

Notes

1. Strains obtained in this way may sometimes revert relatively rapidly. Temperature-sensitive strains and those showing indications of lesions in the membrane may give a higher frequency of stable fragile mutants. Some cell-cycle mutants are fragile only at the restrictive temperature.
2. Fragile mutants determined in the way may be not only temperature-sensitive, but may have increased sensitivity to various drugs such as amphotericin B, rifampicin and others.

REFERENCE

Venkov P, Hadjiolov AA, Battaner E, Schlessinger D (1974) Biochem Biophys Res Commun 56:599–604

Mutants Having Walls More Easily Digested by Lytic Enyzmes (Mehta and Gregory 1980). These mutants were originally isolated with the intention of using them as food yeasts in the belief that the digestibility of the material would be improved.

Required

1. Yeast cells for mutagenesis (48-h cultures).
2. UV lamp and ancillary equipment for irradiation of the cultures (see p. 30–32).
3. Buffers:
   a) 0.1 M Tris-HCl (pH 7.4).
   b) 0.1 M HCl-KCl (pH 2.0) for pepsin.
   c) 0.1 M sodium phosphate buffer (pH 7.5) for trypsin.
   d) 0.1 M Tris-HCl + 0.005 M $MnCl_2$ (pH 8.0) for peptidase.
   e) 0.1 M Tris-HCl; +0.005 M $CaCl_2$ (pH 7.6) for lipase.
   f) 0.1 M Tris-HCl (pH 7.0) for Helicase.
4. 50 ml Erlenmeyer flasks containing 6 ml of YEP-glucose broth.
5. YEP-glucose agar plates.
6. YEP-glucose agar plates containing Helicase (Glusulase; snail enyzme; β-glucuronidase, etc.).
7. Stirring apparatus (Mag-Mix, etc.).

Procedure

1. Mutagenize the cells by irradiation with UV light (1200 and 2400 ergs/mm^2), with cells suspended in Tris-HCl buffer (4 ml in a 60-mm Petri dish, stirred continuously). Survival 2-5%.
2. Transfer each irradiated sample to 6 ml of YEP-glucose broth in a 50 ml Erlenmeyer flask, and incubated at 25 °C for 6-8 h, with shaking.
3. Dilute the samples appropriately (to give 100-150 colonies/plate) and plate on YEP-glucose agar. Incubate 3-4 days.
4. Replica plate the colonies to YEP-glucose agar, YEP-glucose agar + Helicase. Incubate at 37 °C for 3 days.
5. Select colonies growing on YEP-glucose but not on YEP-glucose + Helicase, for further investigation.
6. Test the selected strains for lysis by pepsin, peptidase, lipase (all at 570 µg/ml) and trypsin (1 mg/ml).

REFERENCE

Mehta H, Gregory KF (1981) Appl Env Microbiol 41:992-999

v) Antibiotic-Sensitive Mutants. "Kamikaze" Strains

These mutants are valuable for the study of protein synthesis and related processes in yeast. The yeast cell envelope is impermeable to the many drugs (cycloheximide, emetine, fusidic acid, amicetin, for instance) which inhibit protein synthesizing systems in vitro, and the first step in the study of mutants which are drug-resistant because of alterations in the protein · synthesizing mechanism, is the isolation of strains which are sensitive to these antibiotics. The use of the novel temperature-sensitive "kamikaze" yeast strain BL15 in a selection procedure eliminates the need for screening of large numbers of colonies and the use of large amounts of expensive antibiotics.

Required

1. "Kamikaze" strain BL15, which grows normally at 32 °C but dies at 42 °C, beginning after 40 min exposure to the restrictive temperature.
2. Complete and minimal media, solid and liquid.
3. EMS, thiosulfate solution for neutralization of the mutagen, and other apparatus for use in mutagenesis.
4. Emetine, trimethoprim, other antibiotics having similar effects.
5. YEP-glucose plates containing the antibiotic(s).

Procedure

1. Mutagenize the culture with EMS according to procedure given previously.
2. Dilute the culture 1:5 in liquid YEP-glucose and incubate overnight at 30 °C.
3. Dilute culture to 3.9×10^7 cells/ml approximately, in YEP-glucose + emetine-HCl, 100 µg/ml, and incubate for 1 h at 32 °C.
4. Shift temperature to 42 °C for 7 h (reduces viable cell count from 4×10^7 to 7×10^5 cells/ml).
5. Wash cells with water, twice, to remove drug.
6. Plate on YEP-glucose plates.
7. Either pick colonies to YEP-glucose plates (800 colonies, 100 colonies/plate), and after incubation, replica plate to YEP-glucose plates containing 500 µg/ml of emetine HCl, or replica plate directly to emetine-containing medium.
8. Select colonies which grow on YEP-glucose medium and do not grow on YEP-glucose + emetine medium.
9. Trimethoprim-sensitive strains may be isolated by a similar procedure. The colonies were tested on minimal medium containing 300 µg/ml of trimethoprim.

Note

Emetine sensitivity results from two mutations, trimethoprim sensitivity, from a single-site mutation.

REFERENCE

Littlewood Barbara S (1972) A method for obtaining antibiotic-sensitive mutants in *Saccharomyces cerevisiae*. Genetics 71:305–308

vi) PEP4 Mutants (Zubenko et al. 1982)

These mutations are recessive and pleiotropic, causing deficiencies in carboxypeptidase Y, proteinase A, RNAse(s) and the repressible, non-specific alkaline phosphatase. The function of the PEP4 gene product is apparently required for the maturation of at least five inactive precursors of vacuolar hydrolases, and is thus of interest in investigation of mechanisms for maturation of vacuolar enzymes and possible may serve as a model for the maturation of other yeast enzymes.

Required

1. Yeast strain (in this example, a trp1 deletion, X2180-1B, a gal2).
2. EMS and ancillary materials and apparatus for mutagenesis.
3. Reagents for testing for the ability to cleave the carboxypeptidase Y substrate, N-acetylphenylalanine-β-naphthyl ester (APE); fast garnet GBC, a diazonium salt, which gives a red colour reaction with β-naphthol, the cleavage product of the reaction.
4. Hide powder azure, for detection of proteinase B activity, and hence of proteinase B-deficient mutants.
5. Complete (YEP-glucose), minimal and minimal-phosphate media.

Procedure

1. Mutagenize the yeast with EMS according to previous procedures.
2. Dilute treated cell suspension, plate on YEP-glucose medium, and incubate at 30 °C for 2–3 days.
3. Detect PEP4 mutants by applying an agar overlay containing APE.
4. *prb* mutants were detected by use of the hide powder azure procedure (Zubenko et al. 1979). Proteinase B-deficient mutants fail to release the blue color from the preparation.

Note

When mutants carrying the pep4 mutation are crossed with strains which are wild type (PEP4) for this mutation, and the resulting diploids are sporulated and the asci dissected, the character segregates 2:2. The two wild-type clones are red, but the *pep*4 clones are red, pink or sectored red and yellow, the mutant phenotype being yellow. There is a long phenotypic lag in expression of the mutant phenotype.

REFERENCE

Zubenko GS, Park Frances J, Jones Elizabeth W (1982) Genetic properties of mutations at the *PEP*4 locus in *Saccharomyces cerevisiae*. Genetics 102:679–690

vii) Membranes. Fatty Acid and Inositol-Requiring Mutants

Inositol-Requiring Mutants (Culbertson and Henry 1975). Inositol is a precursor of several membrane phospholipids, which play an essential role in yeast metabolism and growth, since yeasts which are unable to synthesize them die rapidly.

Required

1. EMS and ancillary materials for mutagenesis (p. 36).
2. Haploid yeast strain having the desired auxotrophic or other markers, grown for 48 h in YEP-glucose medium.
3. YEP-glucose agar plates, minimal medium with and without inositol. When added, inositol is used at 2 mg/l.
4. Apparatus for replica plating.

Procedure

1. Mutagenize the yeast according to standard procedures, using a 70-min exposure, which should give 90% kills.
2. Dilute suspension, plate on YEP-glucose agar.
3. Incubate at 30 °C for 3-5 days.
4. Replica plate to synthetic medium with and without inositol. Select colonies growing only on the supplemented medium.

Inositol-Excreting Mutants

Required

1. As shown above in (a).
2. Indicator strains, auxotrophic for inositol and adenine.

Procedure

1. Mutagenize strain or strains with EMS as before, dilute, spread on YEP-glucose plates and incubate at 30 °C for 3-5 days.
2. Spread a lawn of the indicator strain on minimal medium, containing adenine but without inositol, and dry.
3. Replica plate colonies from the YEP-glucose plate to the plates with the lawn.
4. Inositol-excreting mutant colonies are white, surrounded by a red halo, formed by the cells of the inositol-requiring adenine mutant.

Note

Among the inositol-overproducing mutants isolated by this method, the authors reported one which was a non-constitutive mutant in which the mutation apparently affected the synthesis of phosphatidylcholine. The others were mutants constitutive for inositol-1-phosphate synthetase.

REFERENCES

Culbertson MR, Henry Susan A (1975) Inositol-requiring mutants of *Saccharomyces cerevisiae*. Genetics 80:23–40

Greenberg Miriam L, Reiner B, Henry Susan A (1982) Regulatory mutations of inositol biosynthesis in yeast: Isolation of inositol-excreting mutants. Genetics 100:19–33

Sterol-Requiring Mutants. These have been isolated by two methods: first by isolating mutants resistant to nystatin at normal temperatures and in normal medium, but temperature-sensitive at 36 °C, and second, by isolating mutants which were resistant to nystatin in the presence of cholesterol at 22 °C.

Required

1. Complete medium (YEP-glucose), containing ergosterol or cholesterol, 40 mg/l and nystatin, 15–28 mg/l.
2. Wild-type haploid cells, 48-h culture in YEP-glucose.
3. Apparatus for UV-irradiation.

Procedure

1. Plate the cells on YEP-glucose agar + cholesterol, at 10^7 cells/plate, and irradiate with a UV germicidal lamp (30 s at 45 cm distance).
2. Incubate the plates at 22 °C for 2 days.
3. Harvest the cells and plate at 10^6 cells/plate on YEP-glucose agar containing cholesterol and nystatin, and incubate for 2 weeks at 22 °C.
4. Replica plate colonies resistant to nystatin to YEP-glucose plates and incubate 24 h at 36 °C to identify thermosensitive mutants.
5. Test strains resistant to nystatin at 22 °C and unable to grow at 36 °C for ergosterol and/or cholesterol requirements at 22°, 25°, 28°, 30° and 36 °C.

Notes

1. In the original experiments (Karst and Lacroute 1977), 120 such isolates were obtained, 12 of which grew better in the presence of ergosterol.
2. Mutants in the sterol biosynthetic pathway can also be obtained by selection of strains requiring supplementation of the medium with heme for growth. These strains grow poorly at first on medium in which ergosterol or cholesterol was supplied in place of α-aminolevulinic acid. Spontaneous mutants in the sterol biosynthetic pathway arose after a number of generations (Taylor and Parks 1980).
3. Mutants requiring unsaturated fatty acids can be isolated by normal methods, by replica plating mutagenized yeast cells on media with and without

various fatty acids, which are usually incorporated into the medium as the esters.

REFERENCES

Karst F, Lacroute F (1977) Molec Gen Genet 154:269–277
Taylor FR, Parks LW (1980) Biochem Biophys Res Commun 95:1437–1445

viii) Mutants Auxotrophic for 2'-Deoxythymidine 5'-Monophosphate

Normal wild-type yeast cells are not permeable to thymidine monophosphate (dTMP), so that if mutants requiring dTMP are to be constructed, for studies, say, on DNA replication using the incorporation of the analogue 5-bromodeoxyuridine 5-monophosphate as a marker, then mutants which are permeable to dTMP must be obtained (Little and Haynes 1979).

Required: For Mutants Permeable to dTMP

1. Haploid yeast strain, wild type.
2. Media (Minimal medium containing either glucose (SD), 2%, or glycerol (SG), 3% as sole carbon source, containing sulfanilamide (5 mg/ml) and aminopterin (100 μg/ml), vitamin-free casamino acids (0.15%), adenine (30 μg/ml, as the sulfate) and dTMP (100 and 10 μg/ml)).
3. Phosphate buffer.

Procedure

1. Grow cells of the desired yeast strain to exponential phase in YEP-glucose broth, wash in phosphate buffer and plate on the SD and SG medium, at 2×10^7 cells/plate, and incubate at 34 °C for approximately 4 days.
2. Pick emergent colonies, purify and test. Discard petites. These amount to about 92% of colonies on the SD plates.
3. The remaining grande colonies, from the SG plates, are used to obtain mutants which are highly permeable to dTMP. Grow up the primary colonies and plate on SG medium containing aminopterin, sulfanilamide and dTMP (10 μg/ml).
4. Pick and purify emergent colonies. These should be grandes and highly permeable to dTMP.

Required: For Mutants Requiring dTMP

1. dTMP-permeable strains from previous procedure.
2. EMS and ancillary apparatus and reagents for mutagenesis.

3. Phosphate buffer, 0.1 M.
4. Media: YEP-glucose, containing dTMP (100 μg/ml) or SD medium containing 25 μg/ml of dTMP, and YEP-glucose and SD media without dTMP.

Procedure

1. Mutagenize the dTMP-requiring strains (24-48-h culture) with EMS according to standard procedures.
2. Dilute the mutagenized cells in phosphate buffer and plate on YEP-glucose agar containing dTMP or SD agar + dTMP, and incubate 5 days at 30 °C.
3. Replica plate the emergent colonies to YEP-glucose and/or SD medium, with and without dTMP.
4. Incubate the plates for 3 days and isolate clones requiring dTMP.

Notes

1. The frequency of dTMP-requiring mutants is quite low, in the range of 4-11 colonies/11,00 colonies scored.
2. At least one mutation in the PHO80 gene, controlling phosphatase expression, is allelic to one or more mutations conferring permeability to dTMP.

REFERENCE

Little JG, Haynes RH (1979) Isolation and characterization of yeast mutants auxotrophic for 2'-deoxythymidine 5'-monophosphate. Molec Gen Genet 168:141-151

ix) Glycolytic Cycle Mutants

Hexokinase, alcohol dehydrogenase, pyruvate carboxylase, PEP-carboxykinase, phosphofructokinase, mutants in galactose utilization, phosphogluconate dehydrogenase mutants. Examples of the isolation of some of these mutants follow.

Alcohol Dehydrogenase Mutants. These mutants are selected on the basis of resistance to the effects of allyl alcohol. In normal yeasts, allyl alcohol is converted to the toxic compound acrolein and the cells are killed.

Required

Method (a) on Solid Medium:
1. Mutagen (EMS, nitrosoguanidine). Ancillary apparatus and material for mutagenesis of yeast.

2. Yeast strain, cytoplasmic petite, haploid. A petite of any desired strain can be produced by methods described previously.
3. YEP-glucose medium, containing allyl alcohol, filter sterilized (added after autoclaving), 20, 40, 60, 100 mM.

Procedure

1. Grow the yeast strain and mutagenize by standard methods, and spread aliquots containing approximately 10^7 cells on the plates containing allyl alcohol.
2. Pick colonies and spread cells on plates containing the next-highest concentration of allyl alcohol, until mutants resistant to the highest concentration are obtained.

Required: Method (b), in Continuous Culture

1. Continuous culture apparatus (turbidostat).
2. Petite yeast strain, as in method (a).

Procedure

1. Establish the culture in the turbidostat, doubling time approximately 2 h, without aeration.
2. Add allyl alcohol to the medium reservoir at the lowest concentration in the series (12 mM).
3. When the growth rate is restored to the previous level, increase the concentration of allyl alcohol, to approximately double the previous one.
4. Continue to raise the concentration of allyl alcohol incrementally, allowing the growth rate to return to the original level before increasing the concentration again.
5. When the culture is resistant to the desired level of allyl alcohol, sample the fermentor and isolate resistant strains.
6. Make extracts of cells, grown up from the resistant clones in YEP-glucose medium, and determine changes in the enzyme electrophoretically using PAGE-SDS electrophoresis.

Note

Cell extracts may be made on small amounts of cells, mixed with ballotini beads in buffer, and vortexed in a culture tube on a vortex mixer.

REFERENCES

Ciriacy M (1975) Genetics of alcohol dehydrogenase in *Saccharomyces cerevisiae*. I. Isolation and genetic analysis of *adh* mutants. Mutation Res 29:315–326

Wills C, Phelps Julia (1975) Arch Biochem Biophys 167:627–637

Pyruvate Carboxylase Mutants. These mutations affect the anapleurotic reactions which supply intermediates to the Krebs cycle reactions, and such mutants may illuminate the interface between reactions taking place within the mitochondria and those occurring in the cytoplasm. The mutation leading to pyruvate carboxylase deficiency *(pyc)* is a recessive single-gene nuclear mutation. Like biotin-deficient mutants, this mutant has an aspartic acid requirement; pyruvate carboxylase is a biotin-requiring enzyme. Because of the nutritional complexity of the *pyc* mutant, it was constructed in a strain which was otherwise wild type.

Required

1. Appropriate yeast strain (haploid, wild-type), grown in YEP-glucose broth.
2. Media: (a) yeast extract 1%, peptone 1%, plus 2% of glucose, ethanol, potassium acetate, sodium pyruvate, glycerol or DL-sodium lactate (mostly non-fermentable).
   (b) Yeast-nitrogen base (Difco), w/o amino acids, 0.75%, + 2% glucose, ethanol, potassium acetate, sodium pyruvate (1%) or DL-lactate. Two or three hundred plates will be required.
   The above media are solidified with 2% agar if required.
3. EMS as mutagen, and ancillary apparatus and materials for mutagenesis.

Procedure

1. Mutagenize the yeast strain according to standard procedures, 1 h in 3% EMS, in YEP-glucose broth.
2. Recover the cells by centrifugation, resuspend in 6% $Na_2S_2O_3$ for 15 min, recover and wash twice in 50 mM phosphate buffer, pH 7.0.
3. Dilute the cell suspension (original concentration 5×10^5 viable cells/ml), and plate to give 100–200 colonies/plate, on YNB-glucose (100 plates). This step eliminates amino acid auxotrophs.
4. Replica plate these plates (velvet or filter paper) to YEP-ethanol and YNB-ethanol.
5. Isolate colonies growing on YEP-ethanol but not on YNB-ethanol.
6. Mate the mutant to a wild-type strain and isolate segregants unable to grow on YNB-ethanol. Again cross these to wild-type, until strains giving high spore germination ($> 95\%$) and 2:2 segregation of the character are obtained.
7. Cross the mutant to an ADH-negative strain, to demonstrate that the mutant has intact Krebs cycle and oxidative phosphorylation systems (ADH-negative strains do not survive when these pathways are inhibited).

REFERENCE

Wills C, Melham T (1985) Arch Biochem Biophys 236:782–791

Mutants Defective in Phosphoenolpyruvate Carboxykinase (PEPCK). These mutants are selected from mutagenized cultures forming colonies on media containing glycerol as sole carbon source, but not on ethanol. The colonies were isolated and analyzed for the presence of PEPCK.

Required

1. Glassware and reagents, etc., for mutagenesis of yeast with EMS.
2. Suitable haploid yeast strain(s) for mutagenesis and crossing.
3. Media: YEP-glucose, YEP-glycerol, YEP-ethanol, YEP-ethanol + 3 μg/ml of nystatin (for enrichment).
4. Reagents and apparatus for cell disintegration (see note above) and assay for PEPCK (PAGE-SDS electrophoresis).
   (a) 20 mM imidazole buffer, pH 7.0, and glass beads (0.5 mm) for breaking yeast cells by vortexing.
   (b) PEPCK assay reaction mixture (in 0.5 ml): 25 μmol imidazole, 25 μmol $NaHCO_3$, 1 μmol $MnCl_2$, 1 μmol ADP, 0.25 μmol NADH, 1 μmol phosphoenolpyruvate, 80 μkat malate dehydrogenase, 40 μkat hexokinase; cell extract.

Procedure

1. Grow cells of the desired strain in YEP-glucose broth, recover and mutagenize with EMS as previously described.
2. Enrich the cells in mutants by growing in ethanol medium + nystatin, 3 μg/ml.
3. Dilute the cell population suitably and plate on YEP-glucose agar (100 to 200 colonies/plate).
4. Incubate the plates for 3-4 days at 25-30 °C, then replica plate to YEP-glycerol and YEP-ethanol media.
5. Isolate colonies growing on glycerol as sole carbon source but not on ethanol.
6. Grow potential mutants in YEP-glucose broth to yield at least 100 mg of cells (wet weight). These are broken by vortexing with glass beads in 0.5 ml of imidazole buffer and the beads, debris and intact cells are removed by centrifugation.
7. Test the extract for PEPCK levels (Hansen et al. 1976).

Note

Mutants lacking fructose-1,6-biophosphates were not isolated in this procedure. These mutants have been isolated using strains impaired in their ability to grow on glucose, but having an active phosphatase (van de Poll et al. 1974).

REFERENCES

Hansen RJ, Hinze H, Holzer H (1976) Anal Biochem 74:576–584
Perea J, Gancedo C (1982) Arch Microbiol 132:141–143
van de Poll KW, Schamhart DHJ (1977) Molec Gen Genet 154:61–66

x) Secretory Mutants, Transport Mutants, etc.

Secretory mutants are those in which an enzyme or other protein is synthesized but is not transported to its normal location, either in the periplasmic space or outside the cell wall. The best-known secretory mutants are those in which invertase and alkaline phosphatase are not transported to their normal location in the periplasmic space, or in which the mating hormones (a and α-factors) are not transported to their extracellular location, causing sterility in the strain. Killer factors are a special case, in which the toxic protein is synthesized under the control of a double-stranded RNA genome enclosed in the capsid of a virus-like particle (VLP), after which it is transported out of the cell and deglycosylated en route. Invertase normally is found in the periplasmic space, in the glycosylated form. The original secretory mutants were isolated from a group of mutant strains, temperature-sensitive for growth and secretion, which accumulated invertase and acid phosphatase within the cell (Class A *sec* mutants). Class B *sec* mutants do not produce active enzymes at the restrictive temperature.

Transport mutants are generally classed as those in which some substance is not transported into the cell from the external medium. Many of them have a phenotype showing resistance to amino acid analogues, heavy metals, or other inhibitors (Cooper 1981). Transport phenomena may play a part in the failure of petite mutants of some strains to grow on certain sugars; galactose is a familiar example (Evans and Wilkie 1976).

As another example, Goldenthal et al. (1983) isolated a mutant of *Saccharomyces cerevisiae*, deficient in maltose transport, by isolating a number of mutants unable to metabolize maltose, and testing cell-free extracts from them. One of these had normal levels of wild-type maltase, and was assumed to have a defect in maltose transport. This system of isolating transport mutants is useful as a general multi-stage technique for isolating mutants lacking enzymes in the glycolytic cycle, as can be seen from the methods previously described, and can be used to isolate mutants in other metabolic pathways as well. It is more laborious than methods in which the mutagenized culture can be tested directly for deficiencies or resistance to some compound, but these cannot always be applied.

Other mutations giving rise to defects in transport of some compound confer resistance to inhibitors of growth, which may be analogues of a normal metabolite. Methionine and L-ethionine, adenine and 4-aminopyrazolopyrimidine, basic amino acids and L-canavanine, are examples of this phenomenon (Cooper 1981).

REFERENCES

Cooper TG (1981) In: Strathern JN, Jones Elizabeth W, Broach JR (eds) The molecular biology of the yeast *Saccharomyces*. Metabolism and gene expression. Cold Spring Harbor Press, Cold Spring Harbor, New York, pp 399–461
Evans IH, Wilkie D (1976) Genet Res (Cambridge) 27:89–93
Goldenthal MJ, Cohen JD, Marmur J (1983) Curr Genet 7:195–199
Novick P, Schekman R (1979) Proc Natl Acad Sci USA 76:1858–1862
Novick P, Field C, Schekman R (1980) Cell 21:205–215

7. Mapping and Fusion

Mapping of mutant and other genes in yeast to sites on a particular chromosome involves techniques which have been described already in this volume. Most of the procedure involves crossing the yeast strain carrying the unmapped gene(s) to a strain carrying mapped markers. The general approach includes tetrad analysis of the various crosses, random spore analysis if tetrad analysis is not possible, trisomic analysis (for assigning a gene to a particular chromosome), and mitotic mapping methods including mitotic crossing-over (induced by UV light or ionizing radiation or alkylating agents) and mitotic gene conversion, and chromosome loss and chromosome transfer. Fine-structure mapping is used for intragenic maps, and is done by X-ray treatment and treatment with recombinogenic agents, followed by meiotic and mitotic analysis, or, preferably, by deletion mapping.

a) Tetrad Analysis

This method can be used to determine the distance (in centimorgans) between two markers on the same chromosome. If there are no crossovers between the markers, only parental ditypes (PD; AB AB ab ab) asci will be found. A single crossover event will yield tetratype (T; AB Ab aB ab) asci only, and additional crossovers will result in the occurrence of all 3 types of asci, including non-parental ditype (NPD; Ab Ab aB aB). Map distance between the markers is defined as the average number of crossovers per chromatid, x100. In terms of numbers of PD, NPD and T asci, assuming that only 0, 1 or 2 crossovers occur in the interval between the markers, the equation $X = 100/$

$2[(T+6NPD)/(PD+NPD+T)]$ gives the map distance between the markers to within a few percent, if the map distance is less than 35-40 cM. At greater distances, the equation underestimates the true map distance, because of the increasing frequency of triple- and higher-order crossovers in the region.

b) Determination of Centromere Linkage

This is dependent on second-division segregation (SDS) of the parental alleles in meiosis, SDS frequency being zero at the centromere and increasing to 67% for genes located at a long distance from the centromere (40-60 cM) on the chromosome arm. Thus a gene is considered to be centromere-linked if the SDS frequency is significantly less than 67%.

Three methods, two of which are seldom useable, may be used to determine whether a gene is centromere-linked.

a) If a hybrid forms linear asci predominantly, centromere-linked genes segregate (usually) in an alternate array (AaAa) if the asci are dissected in such a way that the spores can be removed from the ascus in order (Hawthorne 1955), while genes which are not linked to the centromere also yield asci having other segregational arrays (AAaa, AaaA, aAAa). Dissection requires a different technique than usual, and is more difficult.

b) If the hybrid is tetraploid and is in a duplex state (A/A/a/a), it yields three types of asci: Aa,Aa,Aa,Aa; AA,Aa,Aa,aa; and AA,AA,aa,aa. For non-centromere-linked genes, the ratio of the three classes of asci are 4/9:4/9:1/0, and for centromere-linked genes, 2/3:0:1/3 (Opportunities for use of these two methods are relatively rare).

c) In diploid strains, producing unordered asci, centromere linkage is determined by including a gene (*trp*1, *pet*8, *met*14, for instance), closely linked to the centromere (SDS less than 1%), and the frequency of T asci is determined, which corresponds approximately to the SDS frequency of that gene, and the map distance of a gene from its centromere is half the SDS frequency, for low values of SDS (corresponding to approximately 10 cM).

Note

Dissection of linear asci, to determine linkage of a gene to its centromere, cannot be done by a method involving enzymatic treatment of the ascus wall, as it is normally not possible to maintain the order of the spores during separation. The ascus can be moved from a drop of suspension to a clear area of the slide, and rolled around and squeezed by the end of the needle, until the spores are extruded from the end of the ascus, one at a time and separated. This method is very slow.

c) Assignment of a Gene to a Particular Chromosome: Trisomic Analysis

i) Known Chromosome is Disomic

If such a strain is crossed with a haploid strain carrying an unmapped gene, and the hybrid is sporulated and the asci dissected, the segregations of the unmapped gene will be 4:0, 3:1 and 2:2 if the gene is on the disomic chromosome, and 2:2 only, if the gene is on any other chromosome. One or more known markers on the trisomic chromosome should be in the duplex condition (A/A/a).

ii) Chromosome Bearing the Unmapped Gene is Disomic

This strain is then crossed to strains carrying markers on all chromosomes. The chromosome carrying the unmapped gene is determined in much the same way as previously.

Note

Disomic strains are usually isolated by recovering them as rare meiotic products of diploid strains showing aberrant phenotypes.

iii) Multiply Disomic Strains

These are often obtained by sporulation of triploids, and isolation of any viable spore clones, which often have two or more (up to 5) disomic chromosomes. The haploid, carrying multiple markers on many chromosomes, plus one unmapped gene, is crossed to these multiply disomic stains. Markers present on disomic chromosomes segregate 4:0, 3:1 and 2:2, and markers on other chromosomes segregate 2:2 only. The unmapped gene can then be excluded from chromosomes that segregate differently from the one where the unmapped gene is located. Up to four to five chromosomes per cross can be eliminated as sites for the unmapped gene (Mortimer and Hawthorne 1973).

iv) Super-Triploid Method (Wickner 1979)

The triploid is constructed with at least one marker on each chromosome in the duplex (A/A/a) condition. The triploid is then sporulated, random spores are isolated and mated with haploid cells carrying the unmapped gene. Several crosses are analyzed genetically, and each cross normally serves to exclude 4.2 chromosomes as the site of the unmapped gene.

d) Mitotic Mapping Techniques

i) Mitotic Crossing-Over

The frequency of this phenomenon is increased from 10^{-4} (spontaneous) to 10^{-2} or 10^{-1}. Mutations such as *rad*18 and *chl*1 will likewise increase the frequency of mitotic crossing-over. Thus if a diploid is heterozygous for several normal markers and an unmapped gene, and undergoes mitotic crossing-over, induced by any suitable method, the markers will be expressed in the sectored colonies, whenever a marker is distal to a crossover event. If the unmapped gene sectors with a known marker, the two are on the same arm of the chromosome. Genes controlling color, colonial morphology, nutritional markers (using omission media), and genes conferring resistance to various toxic agents, can be located on specific chromosomes by this method.

ii) Mitotic Chromosome Loss

This method is based on concomitant expression of two recessive markers in a diploid strain containing an unmarked gene and several known markers, which is taken as an indication that the two are located on the same chromosome. Such diploids in which chromosome loss is induced by recessive mutations such as *cdc*6, *cdc*14, and *chl*1, must be homozygous for this mutation. The mutations *spo*11 and *rad*52 also cause increased rates of chromosome loss. After the unmapped gene is located on a particular chromosome, mitotic recombination and tetrad analysis, or similar techniques, are used to confirm the assignment of the gene to a particular chromosome, locate it on a specific chromosome arm, and map it to a definite site on that chromosome arm.

Notes

1. The mutation *chl*1 causes spontaneous loss of chromosomes I, III, ii, Vii, IX, XII, XV and XVI, and also causes a 10x increase in mitotic recombination.
2. Mutations *cdc*6 and *cdc*14 (ts), in diploids homozygous for either mutation, are heat-shocked (6 h at 36 °C) and undergo marker loss.
3. Chromosome loss induced by methyl benzimidazol-2-yl carbamate (MBC), a benomyl derivative, ρ-fluorophenylalanine, $CdCl_2$, $CoCl_2$, acriflavine: MBC, unlike the other compounds, induces chromosome loss without increasing rates of mitotic recombination or mutation. Chromosome loss occurs in up to 50% of the cells, and most chromosomes are lost at equally frequencies. The compound has potential uses in mapping by chromosome loss.

REFERENCES

Mortimer RK, Hawthorne DC (1973) Genetic mapping in *Saccharomyces*. IV. Mapping of temperature-sensitive genes and use of disomic strains in localizing genes. Genetics 73:33–54

Wickner RB (1979) Mapping chromosomal genes of *Saccharomyces cerevisiae* using an improved genetic mapping method. Genetics 92:803–821

e) Mapping by Chromosome Transfer

This method has not to our knowledge been used as yet for mapping. It depends on the use of haploids carrying the *kar*1 mutation, which prevents or greatly delays nuclear fusion (Conde and Fink 1976), but apparently permits transfer of single chromosomes from one nucleus to the other (Dutcher 1980; Kielland-Brandt et al. 1983), and marked with a gene for canavanine (or other similar drug) resistance, and having each chromosome marked with at least one recessive marker. These strains could be crossed with a near-wild-type haploid containing the unmapped gene. Concomitant appearance of a dominant marker with the recessive marker being mapped (both from the nonselected nucleus) would suggest that both genes are located on the same chromosome.

REFERENCES

Conde J, Fink GR (1976) Proc Natl Acad Sci USA 73:3651–3655

Dutcher Susan K (1981) Molec Cell Biol 1:245–253

Kielland-Brandt MC, Nilsson-Tillgren T, Petersen JGL, Holmberg S, Gjermansen C (1983) Approaches to the genetic analysis and breeding of brewer's yeast. In: Spencer JFT, Spencer Dorothy M, Smith ARW (eds) Yeast genetics: fundamental and applied aspects. Springer, Berlin Heidelberg New York, pp 421–437

f) Fine-Structure (Intragenic) Mapping

This is done by constructing heteroallelic diploids containing the gene to be mapped, and determining (a) the number of prototrophs produced after treatment of the diploid with a recombinogenic agent (UV, X-rays, EMS, MMS; see section on methods for mutagenesis) or (b) the frequency of prototrophic spores among the meiotic products of such diploids, mutagenized as above.

Deletion mapping is nevertheless the method of choice for fine-structure mapping (Fink and Styles 1974).

These methods probably will both be replaced by mapping by gene cloning and DNA sequencing (Fogel et al. 1983; Struhl and Davis 1977; Section 2, this volume).

REFERENCES

Fink GR, Styles C (1974) Genetics 77:231
Fogel S, Mortimer RK, Lusnak K (1983) Meiotic gene conversion in yeast: molecular and experimental perspectives. In: Spencer JFT, Spencer Dorothy M, Smith ARW (eds) Yeast genetics: fundamental and applied aspects. Springer, Berlin Heidelberg New York, pp 65–107
Struhl K, Davis RW (1977) Proc Natl Acad Sci USA 74:5255

g) Strategies for Mapping According to the Above Methods

Method 1

1. Cross the mutant, carrying the unmapped gene, to a strain carrying a centromere-linked marker (*trp*1, for instance). Determine frequency of tetratypes in relation to *trp*1.

2. If this value is less than 2/3 (67%), then cross the original strain to a set of tester strains carrying centromere-linked markers, and look for linkage to one of the markers. If there is none, then the unmapped gene is on a new chromosome (or the centromere of chromosome XVII).

3. When the gene has been assigned to a potential chromosome, do mitotic crossover analysis as above, to assign the gene to a particular arm of the chromosome.

4. Localize the gene to a specific site on the chromosome by tetrad analysis involving markers (preferably more than one) already mapped on this arm.

Method 2

1. Acquire a set of strains carrying 66 markers spaced approximately 50 cM apart over the entire yeast genome.

2. Cross the haploid carrying the unmapped gene with the tester strains and do tetrad analysis on the resulting hybrids.

3. Examination of four to five crosses should enable the mapping of most genes. Genes which have not been mapped successfully by this method are probably located distal to any of the markers or in some other unmarked regions.

Notes

1. Mapping of genes should probably be undertaken only after some experience has been gained in genetic manipulation of yeast, including use of the micromanipulator.

2. Tester strains mentioned above, including the set of 9 strains carrying 66 markers used in strategy 2, can be obtained from the Yeast Genetics Stock Center, Donner Laboratory, Department of Genetics, University of California, Berkeley, California 94720, USA.

GENERAL REFERENCE

Mortimer RK, Schild D (1981) Genetic mapping in *Saccharomyces cerevisiae*. In: Strathern JN, Jones EW, Broach JR (eds) The molecular biology of the yeast *Saccharomyces*: life cycle and inheritance. Cold Spring Harbor Laboratory Press, Cold Spring Harbor, New York, pp 11–26

8. Protoplast Formation and Fusion

a) Regeneration in Solid Medium

Protoplasts can be used in numerous studies, other than for hybridization, including investigations involving isolation of nuclear and mitochondrial DNA, and especially for isolation and separation of relatively intact chromosomes or chromosome-sized sections of DNA. Any procedure in which the presence of the cell wall is an impediment is facilitated by preliminary removal of the wall; i.e. formation of protoplasts.

In addition, fusion of yeast protoplasts is becoming a recognized technique for production of hybrids which cannot be obtained by other means. Interspecific and intergeneric hybrids of yeast species of various types are relatively easy to obtain, though the mechanism of formation is not as simple as was originally hoped, especially by users desiring a rapid and easy method for obtaining new strains of industrial yeasts. The behaviour of the nuclei in an intergeneric, and probably in an interspecific, fusion hybrid appears to resemble more closely the behaviour of the nuclei in fused plant protoplasts (Wetter 1977). One nucleus becomes dominant, and most of the chromosomes from the other nucleus are lost. Some genes are retained, at least in interspecific and intergeneric fusion hybrids of yeasts, and may be integrated into the final genome by the same mechanism as in single-chromosome transfer in normal matings in strains, one of which carries the *kar*1 mutation. However, this mechanism is not yet fully understood, nor is its application to fusion hybrids at all certain. It is known that whatever markers are used to identify the fusion hybrids (auxotrophic nuclear markers or fusion of grandes of one species with a petite of a strain of *Saccharomyces cerevisiae*), the end result is essentially a cell whose genome is principally that of one parent, plus a few genes from the other. This is an advantage in constructing improved industrial yeast strains, as the result is essentially the original industrial strain which has been more or less satisfactory for the user's purpose, plus a few desirable genes from the other species, conferring additional desirable or essential properties. We have recently observed extensive complementation of auxotrophic requirements in *Saccharomyces cerevisiae* diploids, fused with *Hansenula* species (Reynolds et al. 1986, unpublished data).

Protoplasts are also used in the earlier procedures for transformation of yeasts, and fusion routinely occurs during the process. It may be essential for the transfer of the plasmid DNA across the membranes. Newer methods for transformation use cells treated with alkali or lithium acetate, for instance, and fusion of treated cells has not been shown to take place (see section on use of recombinant DNA techniques in yeast genetics).

The University of Washington protocol for formation of yeast *(Saccharomyces cerevisiae)* protoplasts is as follows:

Required

1. Yeast strain, grown to mid-log phase, in YEP-glucose medium (approximately 16-24 h).
2. Solutions: Buffer-pretreatment: EDTA 0.01M; Tris-buffer 0.1 M, KCl 0.6 M (1-2 M if osmotolerant species such as *Zygosaccharomyces rouxii* or *Zygosaccharomyces bailii* are to be used), mercaptoethanol 0.1 M (thioglycollic acid or dithiothreitol may be used.
3. 0.6 M KCl solution, sterile (for fusion as well as protoplasting. 1.2 M sorbitol may be used and may be superior for some applications).
4. Protoplasting buffer, sterile (0.016 M citric acid + 0.016 M KH_2PO_4, in 0.6 M KCl solution).
5. Enzymes for removing cell wall (Glusulase or Helicase, Zymolyase, Novozyme or other. All are effective).
   For Fusion (Ferenczy, pers. commun.)
6. 0.6 M KCl.
7. 0.4 and 0.8 M $CaCl_2$, sterile.
8. Water agar, 3%.
9. Polyethylene glycol, 4500-6000 (PEG), 35%, containing 15% dimethylsulfoxide (DMSO).
10. Osmotically stabilized (with 0.6 M KCl) YEP-glucose agar plates, for regeneration of fused protoplasts (plates should be prewarmed before use), or:
11. OSY-glucose broth, containing 35% PEG, for regeneration in liquid medium, if desired.
12. Sterile tubes, one for each regeneration plate desired.
13. Water baths, set at 36-37° and 44-45 °C respectively.

14. Sterile 1- and 10-ml pipettes; the latter large bore. Plastic disposable pipettes are satisfactory.

Procedure

1. Spin down cells from 2-10 ml of 16-24-h culture (approximately 2×10^8 cells).
2. Wash cells once if desired.
3. Add pretreatment solution (4 ml), resuspend cells, and incubate at 37 °C for 10-15 min, depending on the strain and species.
4. Spin down and wash once with sterile 0.6 M KCl.
5. Spin down again and resuspend in protoplasting solution (2 ml).
6. Add 0.1-0.2 ml of enzyme solution, or if Novozyme is used, make the final concentration of enzyme 1 mg/ml.
7. Incubate at 35-37 °C until protoplast formation is complete (10 to 90 min, usually, depending on the strain). Determine the progress of protoplasting by microscopic observation, and addition of a drop of water at the edge of the slide. Protoplasts should burst. The completeness of protoplasting should be confirmed by plating an aliquot on normal (not osmotically stabilized) YEP-glucose agar. Intact, viable cells will form colonies.
8. During the incubation period, prepare the tubes of overlay agar. Add 2.5 ml of the 0.8 M $CaCl_2$ solution to each tube, and hold at 44-45 °C. To each then add 2.5 ml of molten 3% water agar. Keep tubes at 45 °C.
9. Spin down protoplasts at relatively low speed, and resuspend in 2 ml of sterile 0.4 M $CaCl_2$ solution.
10. Dilute if necessary, after counting in a hemocytometer, then centrifuge an aliquot or the whole sample.
11. Resuspend in 0.4 M $CaCl_2$ and mix approximately equal proportions of the strains to be fused.
12. Centrifuge, decant supernatant, and resuspend mixture in the residual liquid (approximately 0.2 ml).
13. The mixture may be diluted again at this point, and 0.2 ml of the mixture may be taken for fusion. If a less dense suspension is used, the average number of protoplasts in each aggregation is reduced, which may or may not be an advantage.
14. To the final protoplast mixture, add 2 ml of 35% PEG + 15% DMSO solution.
15. Incubate for 10-30 min at 36-37 °C.
16. Dilute the fusion mixture 10x and 100x in PEG or 0.6 M KCl, and add an aliquot (0.1 ml) to a tube of overlay agar. Mix rapidly and *immediately* pour the whole contents of the tube over the surface of a prewarmed plate of OSY-glucose agar.

Protoplast Formation and Fusion

17. When the overlay has set, incubate the plates at 25-30 °C until colonies appear.
18. Pick colonies to selective media; YEP-glycerol to eliminate petites, if any, and other selective agars as determined by the nature of the other parent. If auxotrophic strains are fused, the fusion products may be selected on minimal or dropout media.
19. If most of the colonies have broken through the overlay, then the colonies may be replica-plated (velvet pads or filter paper) to the selective media.
20. Or, the fusion products may be regenerated in liquid medium containing 35% PEG (Svoboda and Piedra 1983).

b) Regeneration in Liquid Medium

Regeneration in liquid medium can, in theory, be used to recover all of the fusion products resulting from PEG-induced fusion of a mixture of protoplasts. Normally, protoplasts of *Saccharomyces cerevisiae* will not regenerate the cell wall in liquid medium, and it has been assumed that they require some kind of solid support, which may be agar, gelatin or other similar materials. However, it has recently been shown that they will regenerate a wall and form normal cells in a liquid medium if PEG is present in concentrations above 30% (S. Darling, pers. commun.). The medium is somewhat tedious to prepare, as the PEG must be sterilized separately, but regeneration in liquid medium should permit the isolation of rare hybrids, produced in such fusions, and which would otherwise be recovered only by chance from platings on solid medium in overlays. Picking all of the colonies obtained if all of the fused material were plated would be extremely tedious at best, and probably impossible as a practical matter, and replica plating does not recover those colonies which remain buried in the agar.

Required

1. YEP-glucose broth, containing 35% PEG. As previously noted, the PEG and YEP-glucose should be made up separately in double strength and mixed after autoclaving. Dispense in 10-ml batches in Petri dishes, or in 250 ml Erlenmeyer flasks, 20 ml/flask.
2. YEP-glucose broth + 0.6 M KCl.
3. Selective media for isolation of hybrids.

Procedure

1. Transfer the fused protoplast mixture, in 0.5 ml aliquots, to the YEP-glucose mixture. It may be desirable to wash the fused protoplasts in PEG

or 0.6 M KCl to remove DMSO from the suspension. Some investigators prefer to wash with 1.2 M sorbitol solution to remove both PEG and DMSO, after fusion.

2. Either
   a) Incubate for 8-12 h at 39 °C.
   b) Remove PEG solution, transfer regenerating fused protoplasts to osmotically-stabilized YEP-glucose broth, and incubate 24-48 h at 30 °C.
   c) Spin down cells, discard supernatant, transfer regenerated cells to selective medium.
   d) Incubate for 24-48 h at 30 °C, in selective media, transferring to other media as required.
   e) Plate out on final selective, solidified media for isolation of hybrid clones.

   Or, omit step (b) above, and incubate for 24-48 h before transfer of regenerated cells to selective media.

Notes

1. It may be possible to substitute compounds such as polyvinyl alcohol (PVA) or carboxymethyl cellulose (CMC) for PEG in the regeneration step, if PEG proves too toxic for some strains.
2. The procedure for interspecific and intergeneric fusions is the same as for intraspecific fusions (Spencer and Spencer 1981; Spencer et al. 1983). If one of the partners in the fusion is a strain of *Saccharomyces cerevisiae*, then use of the petite mutant of the desired strain may make elimination of that parent easier, with consequent increased ease of isolation of the fusion products. It has been shown that mitochondria from some other yeast species (*Hansenula capsulata, Hansenula wingei, Kluyveromyces lactis, Saccharomyces rosei (Torulaspora delbrueckii), Saccharomyces montanus*, and others) can either function as such in *S. cerevisiae*, or the mtDNA can at least recombine with non-functional mtDNA of the petite mutant, if present. In these crosses, acquisition of parts of the nuclear genome of the non-*Saccharomyces* parent appears to take place by single-chromosome transfer or some similar mechanism, rather than by a "merging" of the two genomes. However, when hybrids are isolated by complementation of nuclear markers, one nucleus or the other becomes dominant (Groves and Oliver 1984), as in fusion of plant protoplasts from different species, so that the acquisition of a few characters by single-chromosome transfer is probably as effective in producing new hybrids, and has the advantage that the dominant hybrid isolated, in this case, is essentially the original strain of *Saccharomyces*, which has acquired one or a few desirable characteristics from another species.

3. The percentage regeneration of the protoplasts may sometimes be increased by incubation of the protoplasts in YEP-glucose broth, osmotically stabilized, for 30–40 min, before recovery of the protoplasts, resuspension in 0.4 M $CaCl_2$, and subsequent fusion in PEG.
4. Osmotolerant yeast species yield protoplasts which are fragile in normal concentrations of osmotic stabilizers. Protoplasting of *Zygosaccharomyces rouxii*, for instance, requires 2 M KCl or equivalent, if the protoplasts are to be stable, and similar adjustments must be made in the concentrations of other osmotically-active materials used in the fusion process. Other species requiring higher concentrations of osmotic stabilizers include *Zygosaccharomyces bailii, Zygosaccharomyces cidri* and *Candida mogii*.
5. Fused protoplasts may also be regenerated in YEPG-gelatin containing approximately 0.4% agar, which yields a solid at room temperature (20–25 °C) but melts at 35 °C, after regeneration has taken place, so that the cells can be recovered and placed in elective media for enrichment in the desired hybrids.
6. Isolated mitochondria have been transferred into petite strains of *Saccharomyces* by fusion with mini-protoplasts, containing mitochondria but no nuclei (Fukuda and Kimura 1980). The mini-protoplasts were collected by differential centrifugation, large protoplasts containing nuclei being removed by centrifugation at 1950 g (5 min), after which the mini-protoplasts were collected by centrifugation at 3100 g for 15 min and washed with protoplasting buffer, before fusion in 35% PEG plus 50 mM $CaCl_2$.

REFERENCES

Fukuda H, Kimura A (1980) FEBS Lett 113:58–60
Groves DP, Oliver SG (1984) Curr Genet 8:49
Spencer JFT, Spencer Dorothy M (1982) Curr Genet 4:177–180
Spencer JFT, Spencer Dorothy M, Whittington-Vaughan P, Miller R (1983) Curr Genet 7:159–164
Svoboda A, Piedra D (1983) J Gen Microbiol 129:3371–3378
Wetter LR (1977) Molec Gen Genet 150:231–235

Part II. Methods Using Direct Manipulation of DNA and RNA

1. Separation of Large DNA Molecules, Greater than 25 Kbp, by Gel Electrophoresis

Introduction

Gel electrophoresis is a technique that is used to separate different-sized ions as they migrate through a gel medium in response to an electric current. However, molecules larger than a certain size, be they DNA or protein-detergent complexes, all co-migrate, or effectively so, and thus cannot be characterized by conventional one-dimensional electrophoresis. In the case of DNA separations performed in agarose slab gels, it is possible to lower the concentration of agarose to less than 0.5% to improve separation efficiency, but gels then become very difficult to handle, in fact by using 0.1% agarose gels molecules up to 500 Kbp in size can be separated (Fangman 1978; Serwer 1981). Some characterization of large DNA molecules has previously been carried out by electron microscopical, velocity sedimentation and viscoelasticity studies but little real progress has been made until recently, when new electrophoretic techniques were introduced.

The problem associated with the separation of large DNA molecules (> 25 Kbp) by gel electrophoresis is that their migration through the gel matrix is not strictly related to their size. These large molecules, regardless of size, are believed to assume an 'end-on' configuration to the gel matrix and wriggle through the gel interstitial spaces. This is called reptation (de Gennes 1978; Lerman and Frisch 1982; Lumpkin et al. 1985). With the introduction in 1983 of pulsed field gel electrophoresis (PFG) in agarose gels (Schwartz et al. 1983) and its subsequent development in 1984 to produce orthogonal-field alternation gel electrophoresis (OFAGE) and 1986 to produce field inversion gel electrophoresis (FIGE) (Carle et al. 1986) came the ability to separate DNA fragments of up to 5–10 Mbp. This meant that whole chromosomes from simple organisms could now be separated on the basis of size by gel electrophoresis. Most recently a modification of OFAGE and FIGE has been introduced, contour-clamped homogeneous electric field electrophoresis (CHEF) (Chu et al. 1986).

OFAGE, FIGE and CHEF alter the way large DNA molecules migrate through the gel matrix, imposing a relationship between size and distance migrated. OFAGE systems rely on four electrodes which force the DNA molecules to 'turn corners' as they move through the gel (Carle and Olson 1985) FIGE systems rely on two electrodes and their outward appearance is more like that of a conventional one-dimensional agarose slab gel tank. The heart of both systems is a switching mechanism which alters the direction of the current passing through the gel. In FIGE this can be altered in two directions, forward and backwards, and in OFAGE, a number of directions depending upon the orientation of the four electrodes relative to the gel. It is the frequency of the field inversion that is important in determining the size of the DNA molecules that can be resolved and as the 'gel run' proceeds, the frequency of field inversion can be altered. The whole process can be controlled by a computer program. In OFAGE field shape (dependent on the orientation of the electrodes relative to the gel) is also important in determining separative ability.

Both OFAGE and FIGE systems are currently available commercially. LKB markets the OFAGE 'Pulsaphor' system (Pulsaphor Electrophoresis Unit) Hybaid a FIGE system (Gene-Aid Fige system) and DNA STAR a FIGE switching system (DNA Star). The FIGE system is cheaper, as a standard gel tank can be used, whereas for OFAGE a purpose-built tank must be bought or constructed. Instructions for construction of an OFAGE system are given in Carle and Olson (1984). It has proved useful to use a combination of both OFAGE and FIGE to subject the DNA molecules to two-dimensional electrophoresis to achieve a good separation (Carle et al. 1986). Figure 10 shows a typical FIGE system as used in the author's laboratory.

As previously mentioned, during "steady state" electrophoresis large DNA molecules migrate in a way not strictly related to their size as their configurations within the gel become distorted and they display reptation. It is thought that during reptation the leading segment of the DNA molecule is less in radius than that which follows. This gives rise to a wedge-shaped configuration. The main reason for assuming a wedge shape by large DNA molecules is that the path favored by the molecule to travel through the gel is limited in radius and as a result only a fraction of the molecule can pass through it. Sections of the molecule further from the leading edge penetrate other less favorable paths in the gel matrix, and as a result travel less slowly as forward migration depends upon their extrication from these paths and reorientation along the forward path. This leads to a bunching of the trailing segments of the molecule. When the field is inverted or its orientation relative to the gel changed, it is thought to affect the wedge conformation. This change presumably occurs at one end of the wedge and progresses to the other. Finally, if the field inversion continues for long enough the wedge conformation

Fig. 10. FIGE system used in the author's laboratory (GENE-AID FIGE system) (Parent et al. 1985). *1* Cooling recirculation for buffer; *2* Peristaltic pump for electrolyte cooling and recirculation. *3* Gel tank. *4* Power pack. *5* Switching device. *6* IBM portable computer

would be completely reversed and mobility would begin in the new direction of the electric field. However, if the duration of the field inversion is limited then the wedge conformation never becomes fully inverted. In fact, by controlling the frequency of field inversion the molecule can be kept in an intermediate conformation between one wedge form and another. This intermediate conformation is directly related to the size of the molecule and has a different electrophoretic mobility to the wedge form. The switching frequency is therefore highly important in determining the size of molecule that can be separated.

It has been suggested (Carle et al. 1986) that OFAGE does not present notable advantages over FIGE since better OFAGE separations are achieved when the angles between the applied fields are obtuse; i.e. $> 90°$. It has been postulated that separation by OFAGE could therefore arise from the same head-tail conformational changes that occur in the case of FIGE. If this is the case, then FIGE may eventually replace the more mechanically complicated OFAGE.

The main use to which OFAGE and FIGE have been put is karyotyping simple eukaryotes (Johnston and Mortimer 1986; De Jonge et al. 1986; Carle and Olson 1984). It has also been put to use for examining the antigenic variation and chromosome rearrangements in *Trypanosoma brucei* (Van der Ploeg et al. 1984a,b), and recently pulsed-field electrophoresis has been used to aid in the cloning of large DNA fragments in yeast/*E. coli* shuttle vectors (Burke et al. 1987). In higher eukaryotes, chromosomes are too large to be separated as individual bands by OFAGE or FIGE, the smallest human chromosome having been estimated to be 30 Mbp and the average size to be 150 Mbp. However, by the judicious use of restriction enzymes which recognize relatively rare DNA sequences, such as NotI and SfiI (recognizing eight-base sequences), large fragments can be generated from whole chromosomes, separated by OFAGE or FIGE and probed to identify the location of genes. In this way maps of fragments too large to be separated on an individual basis by OFAGE or FIGE can be generated.

Such techniques are currently being used to map the genomes of higher eukaryotes including man. Recently a 1500 Kbp sequence of the mouse genome has been mapped using the method (Brown and Bird 1986) and PFG has been used in the molecular analysis of the Duchenne muscular dystrophy region of the human X chromosome (Kenwrick et al. 1987), cystic fibrosis (Estivill et al. 1987; Collins et al. 1987), Fragile X syndrome (Patterson et al. 1987) and the construction and use of human chromosome jumping libraries (Collins et al. 1987; Poustka et al. 1987). A recent review of the application of PFG to molecular analysis of human monogenic diseases can be found in (Davies and Robson 1987).

CHEF, a modification of OFAGE and FIGE involving a more complex array of electrodes has recently been developed. It has the capability, unlike OFAGE, of producing a uniform or homogeneous electric field throughout the gel. The inventors claim that using their apparatus allows the comparison of more samples on the same gel and a cleaner, more interpretable appearance to the gel after it has been run. OFAGE tends to produce bands which can be very diffuse or irregular in shape as a result of inhomogeneity of the electric field throughout the gel. This inhomogeneity means that DNA molecules can migrate with different mobilities and trajectories, depending on where in the gel the samples are loaded. This can make comparison of multiple samples on the same gel difficult. New electrode configurations are currently being explored to try to improve the geometry of the DNA tracks and a recent general review of pulsed field gel electrophoresis can be found in Anand (1986). Methods for sample preparation, running gels and construction of apparatus are given in Van Ommen and Verberk (1986).

a) Preparation of Intact Chromosomal-Sized Yeast DNA Molecules

Methods

The preparation of very high molecular weight DNA from yeast cells requires considerable care so that shearing of the molecule does not occur. There are three principal methods which have been developed for the purpose of generating high molecular weight DNA from yeast cells, two of which rely on immobilization of the cells either in agarose beads (Cook 1984) or plugs (Schwartz and Cantor 1984) and third, on recovery of very high molecular weight DNA from sucrose gradients (Olson et al. 1979).

The protocol which follows is a modification of the method given in Schwartz and Cantor (1984), which is technically simpler than the others but gives consistently good results in our hands.

Required

1. EDTA, 0.125 M, pH 7.5.
2. EDTA, 0.05 M, pH 7.5.
3. EDTA, 0.5 M, pH 9.0.
4. EDTA, 0.5 M/0.01 M Tris-HCl, pH 7.5 (LET buffer).
5. EDTA, 0.5 M/0.01 M Tris-HCl/1% sodium lauryl sarcosinate, pH 9.5 (NDS buffer).
6. 2-mercaptoethanol.
7. Proteinase K.
8. Agarose (low melting temperature, LMT).
9. Enzyme(s) for protoplasting (Helicase; Novozyme, Zymolyase).

Procedure

1. Grow cells, with shaking, in 50->200 ml of YEPD medium (liquid) at 30 °C until cell density reaches 10^7-10^8 cells/ml.
2. Pellet in a refrigerated bench-top centrifuge at 2000 g for 5 min at 4 °C and wash twice in 0.05 M EDTA, pH 7.5.

Operations 3-6 should be carried out at 37 °C

3. Resuspend pellet in 1->4 ml of 0.125 M EDTA, pH 7.5, depending on original cell density by agitation, and transfer to sterile capped, plastic disposable centrifuge tube. The suspension should appear as a thick slurry.
4. Add molten (37 °C) 1% LMT agarose in 0.125 M EDTA pH 7.5, to a final concentration of 0.5% and mix well by tapping the base of the centrifuge tube.

5. Add stock solution of 10 mg/ml Novozyme (note 1) in 0.05 M EDTA pH 7.5 to required final concentration (generally with our laboratory strains this is 1 mg/ml final concentration).
6. Add the suspension to holes, precut in 1.5% agarose poured in 6 cm petri dishes, with a depth of 5- >7 mm (note 2).
7. Allow plugs to solidify at room temperature or 4 °C.
8. When set, remove the plug by excision with a scalpel blade slice into small cubes and place in a sterile 6 cm Petri dish and cover with LET buffer (requires approximately 5- >10 ml).
9. Add 2-mercaptoethanol to a final concentration of 7.5% v/v.
10. Incubate Petri dishes at 37 °C for as long as is required for protoplast formation to occur (note 4) (This can be as long as 48 h depending on strain. Protoplast formation can be followed microscopically).
11. Remove agarose plugs to clean Petri dishes and wash with 0.5 M EDTA, pH 9.0.
12. Place agarose plugs in clean Petri dish and cover with a solution of 2 mg/ml of proteinase K in NDS buffer and incubate at 50 °C for 24 h.
13. Remove the NDS buffer and wash the plugs in 0.5 M EDTA, pH 9.0. The agarose plugs can either be stored dry or under NDS buffer at 4 °C in cling film-sealed dishes (note 5).
14. When required, sections of the blocks are cut with a sterile scalpel blade to dimensions required to fit the size of well to be used during electrophoresis.

Notes

1. We normally use 13 ml polycarbonate conical-bottomed screw-capped disposable centrifuge tubes (Sterilin, Middlesex, England, product No. 144ASS).
2. A number of enzymes are commercially available for protoplasting yeast and other fungal cells. These include: β-glucuronidase and zymolyase, both available from Sigma Chemical Company, P.O. Box 14508, St. Louis, Missouri 63178, USA, and Novozyme produced by Novo Industries SA, Copenhagen, Denmark. Fungal cell walls are compared of a mixture of sugars and glycoproteins, including mannans and glucans. The actual composition, chemical structure and linkages found between sugars is strain-specific. Certain of the enzymes are β-1,4-glucanases, while others are α-1,4-glucanases. The choice of enzymes for protoplasting depends on the sugar constituents and linkages occurring in the cell walls of the strain (and species) being investigated. Many of these protoplasting enzymes, especially the commercially-available β-glucuronidases, are impure and contain other enzyme activities including proteases, sulfatases and DNAases.

This can assist in the removal of mannans bound to proteins (glycoproteins) in the yeast cell wall, but care should be taken with respect to suspected contamination with DNAases, which may degrade the yeast DNA. We have found that preincubating the yeast cell pellet in a solution of 0.9 M KCl 0.015 M $MgSO_4$ 0.0125 mM Tris-HCl, pH 7.5 1% v/v 2-mercaptoethanol, before commencing stage 3, can improve the DNA recovery from the agarose plug, probably by bringing about a higher percent of conversion of the cells to protoplasts.
3. An alternative method to plug pouring in agarose-filled Petri dishes has been developed (Johnston and Mortimer 1986) in which plugs or blocks of agar can be formed in disposable plastic cuvettes. The procedure is as follows: Cut off one end of a plastic cuvette with a sharp scalpel blade and seal one end with parafilm. Put the well-mixed cell suspension from stage 5 into the cuvette (you may need more than one cuvette) and allow to solidify at room temperature or 4 °C. When the agarose has solidified, remove the parafilm and push the preformed block out. Place in a sterile 6-cm Petri dish and cover with 0.5 M EDTA, 0.01 M Tris-HCl buffer, pH 7.5. Then continue the procedure, beginning at stage 9. The LKB Pulsaphor comes complete with a block-making mould.
4. The length of time required to produce protoplasts is strain-dependent. It is quite useful to carry out a few preliminary time-course experiments with cells and protoplasting enzyme to gauge how long it will be necessary to leave the agarose plugs before protoplasting will have been completed. It should be kept in mind that yeast cells in the exponential phase of growth are converted to protoplasts much more readily than those in late exponential or stationary phase.
5. An alternative storage method is to place the blocks in a 6-cm Petri dish containing a small 3-cm Petri dish filled with sterile distilled water. When this method is used, there should be no necessity to seal the dish with parafilm, as evaporation from the small Petri dish should keep the agarose blocks/plugs from drying out.

b) Gel Preparation

Required

1. Agarose, Sigma type I low endoosmosis.
2. 0.089 M Tris-borate/0.089 M boric acid/0.002 M EDTA, pH 8.0 (TBE buffer, normally made at 5x strength as stock solution).
3. 0.04 M Tris-acetate/0.001 M EDTA (TAE buffer, normally made up as 50x stock solution).

Method

Agarose is dissolved by heating with 0.5 x TBE or TAE to a final concentration of 1% v/v (This can varied depending on personal requirements. From 1-1.5% has been used in separation of yeast chromosomes). Care must be taken that all the agarose has been dissolved by boiling vigorously for a few minutes. A microwave oven may also be used. Allow the agarose to cool to about 50 °C before pouring into the gel tray, to a depth of approximately 0.5 → 1 cm. Leave to set a room temperature.

The gel dimensions and electrode distances can affect results, and published data refer to specific volts/cm values and gel dimensions to achieve the separations reported (see Running Methods, p. 81).

c) **Size Standards**

The determination of the sizes of large DNA molecules requires suitable marker DNA on the same gel. This marker DNA can be derived in two ways, first, by the use of DNA from bacteriophages such as T4, T2, and G, or, secondly, by the production of λ oligomers. Methods for the isolation and use of bacteriophage DNAs are given in Fangman (1978). The method that follows is for the preparation of agarose blocks containing λ oligomers (Van Ommen and Verkerk 1986). Pharmacia (Molecular Biology Division, Uppsala, Sweden) currently have have under production, λ oligomers, which are supposed to be available in late 1987 (pers. commun.).

Required

1. λ DNA.
2. 0.075 M NaCl/0.025 M EDTA, pH 7.4 (SE buffer).
3. 0.01 M Tris-HCl pH 7.4/0.005 M EDTA (TE buffer).
4. Agarose, (low melting temperature, LMT).

Method

1. Make up λ DNA solution to 20 μg/ml in SE buffer and warm to 37 °C.
2. Add an equal volume of 1% LMT agarose at 37 °C in SE buffer and mix thoroughly.
3. Add to moulds cut in 6 cm agarose-filled Petri dishes or to cuvettes with end removed (see note 3 previous) and allow to solidify.
4. Collect blocks and place in a sterile Petri dish and cover with SE buffer (5->10 ml), incubate at 37 °C for 3-5 h, followed by holding overnight at room temperature.

5. Wash blocks several times in sterile distilled water, followed by several washes and one soak in TE buffer for 5 h.
6. Store blocks at 4 °C in 0.5× TE buffer.
   Oligomeric λ ladders up to approximately 1.5 Mbp can be formed in this way.

Note

Commercial preparations of λ DNA can be unsatisfactory for making λ oligomers. This is probably due to the fact that the cohesive ends have become damaged, either during preparation of the DNA or in subsequent cycles of freezing and thawing caused by storage and use. It is possible to prepare home-made λ oligomers using λ DNA prepared in the laboratory. To do so, λ virions can be embedded in agarose plugs at a DNA concentration of 15–>30 μg/ml. The plug can then be processed as for the embedded yeast cells with the exception that protoplasting is *not* required. Oligomer formation is then allowed to proceed for one to two weeks at room temperature. Half a plug has been found to be sufficient for a gel slot (Anand 1986). The isolation of λ virions is fully described in Maniatis et al. (1982).

A final alternative to the use of λ oligomers is yeast chromosomal DNA. *Saccharomyces cerevisiae* strain X2180-1B chromosomal DNA can be used (Carle and Olson 1984; Anand 1986; Anand, pers. commun.) as size markers (Table 1).

Table 1. Approximate sizes of fractionated chromosomes of *Saccharomyces cerevisiae* strain X2180-1B by pulsed-field gel electrophoresis

Band number	Chromosome number	Length (Kbp)
15	XII	1850
14	XII	1750
13	IV	1370
12	VII, XV	1115
11	XVI	970
10	XIV	940
9	II	820
8	XIV	790
7	X	750
6	XI	680
5	V, VIII	590
4	IX	450
3	III	360
2	VI	280
1	I	245

ND = not determined

d) Restriction Digests

Required

1. Bovine serum albumin (Sigma, molecular biology reagent grade).
2. Restriction enzyme as desired.
3. Restriction enzyme buffer (as specified in manufacturer's data sheet).

Method (modified from those in Anand (1986) and Van Ommen and Verkerk (1986))

1. Place well-sized fragment of block (or chunk required) into a sterile 1.5-ml Eppendorf tube (de Gennes 1978; Lerman and Frisch 1982; Lumpkin et al. 1985).
2. Equilibrate each block with 1.0 ml of the appropriate restriction enzyme reaction buffer overnight at 4 °C or for 2 h at the optimum temperature for enzyme activity (note 2).
3. Remove the reaction buffer with a pasteur pipette and replace with a fresh volume, just sufficient to cover the block and add the restriction enzyme to an approximate final concentration of 2–4 units/μg of chromosomal DNA.
4. Incubate at the desired temperature overnight (note 3).
5. Remove reaction buffer and rinse blocks twice in 0.5 ml of TE buffer.
6. Place blocks into wells or store, covered with TE buffer, in Petri dishes sealed with cling film (note 4).

Notes

1. If the blocks have been stored covered with NDS it is necessary to wash them thoroughly in TE buffer (3 × 30 min) on ice before commencing the restriction digest. Some workers include 0.1 mM PMSF (phenylmethylsulfonyl fluoride), a protease inhibitor, in these washes.
2. and 3. Anand (1986) uses a pre-equilibration buffer prior to digestion, composed of the required reaction buffer minus dithiothreitol (DTT), spermidine and/or gelatin, for 30 min on ice. This is then replaced by a buffer containing all of the necessary constituents plus the restriction enzyme (2–4 units/μg of DNA). The mixture is then incubated for 2 h at the optimum temperature for enzyme activity to produce complete digestion.
3. It has been suggested that blocks should be washed in NDS and digested with proteinase K for 2 h before loading. To ease loading of the blocks into well, they can be stored in TE buffer containing orange-G dye (Anand 1986). This colours the blocks and makes them easier to handle.

e) Conditions for Electrophoresis

i) Loading the Gel

Samples can be loaded in block form or can be melted and allowed to set in situ in the well before running the gel.

Method 1

1. Remove all supernatant liquid from the gel blocks and transfer to the well with a Pasteur pipette, the end of which is bent into a hook. Alternatively use Millipore filter forceps with flat unridged ends, taking care not to squash the blocks. Take care also not to damage the well, as this can cause problems with the migration of DNA into the gel.
2. Push block into well.
3. It may be necessary to set the blocks in the wells with a little melted 1% agarose.

Method 2

1. Remove all supernatant from the gel blocks and transfer to sterile 0.75 ml Eppendorf tubes.
2. Place on heater block or in water bath at 65 °C and melt for 10 min.
3. *Gently* pipette into the wells in the gel using a Gilson pipette with the tip cut off to produce an orifice of diameter greater than 1 mm. (This is to prevent shear).
4. Allow agarose to set in well.

ii) Running the Gel

1. Place gel in the gel tray and cover with buffer 0.5 × TAE or 0.5 × TBE and allow to soak for 30 min.
2. Electrophoresis.

Pulse time, agarose concentration, voltage, temperature and electrode configuration are all important in determining the separative ability of the system being used. Before actual electrophoresis to separate the yeast chromosomal material begins, the DNA can be electrophoresed into the gel by a single unpulsed electrical field. The duration and voltage depend on the size of the gel, but 45 min at 100 V for a 10 cm box and 60 min at 220 V for a 20 cm box have been used (Anand 1986).

f) OFAGE

Carle and Olson (1985) describe a system using a constant voltage of 300 V with a switching time of 50 seconds at 13 °C, applied for 18 h to a 1.5% agarose gel, 10.2 × 10.2 × 0.32 cm, in a tank with electrode lengths and dimensions as published (Carle and Olson 1984), as a general technique for karyotyping a 'standard' *Saccharomyces cerevisiae* strain. Using this regime, they were able to resolve 12 bands on the gel. Three of the bands were doublets, separated by slightly altering the agarose concentration to 1.2% and reducing the switching time to 30 s. Gels were stained in 0.1 μg/ml of ethidium bromide in 0.5 × TBE for 10->30 min, and subsequently destained for 20 min in 0.5 × TBE alone. The size range of chromosomes separated extended from 245 Kbp to in excess of 980 Kbp. The sizes of the three largest chromosomes, VII, XV and IV, could not be determined because of the lack of suitable size markers.

Carle and Olson (1984) describe in detail the effects of pulse time on separation of yeast DNA molecules. Generally the longer the pulse interval, the larger the molecules that can be separated. Temperature can also affect the separative capacity of the gel systems. Generally a temperature near 14 °C is maintained in the gel system by recirculating the gel buffer (the temperature of the buffer tends to rise as both high current and power are used to operate the gels). Snell and Wilkins (1986) have reported that increasing the running temperature of the gel system allows the separation of larger DNA molecules, as the mobility of all DNA molecules is increased. By using temperatures of 25 °C and above, they successfully resolved the largest chromosomal DNA bands of *S. cerevisiae*. They found, however, that resolution fell off at temperatures above 35 °C.

g) FIGE (Fig. 10)

The procedure for the separation of yeast chromosomal DNA molecules ranging from 14 kbp to > 700 kbp in size are given in Carle et al. (1986). Carle et al. (1986) discovered that simple fixed frequency field inversion was not sufficient to produce a good separation of the molecules, but that 'ramping' of the pulse time was necessary. Ramping is the alteration, with increasing time of the run, of the duration of the switching interval or field inversion. In their experiments, the forward switching interval varied from 9 s at t = 0 h to 60 s at t = 18 h. The reverse interval was either 1/4 or 1/3 of the duration of the forward interval. The gels were run at 13 °C for 18 h at a voltage of 300 V. The gels were 21.5 × 20.5 × 0.4 cm in size, of 1% agarose and the voltage gradient between the electrodes was 10.5 V/cm. Carle et al. (1986)

also experimented with a variable forward pulse interval (10 s at t = 0 h to 60 s at t = 12 h), but with a constant reverse pulse interval of 5 s. This produced reasonable results in separating chromosomal-sized yeast DNA molecules.

Recently it has been suggested that a pause interval between the forward and reverse switching can improve separation and that it is better to use a percentage of the forward switching interval as a pause rather than a fixed time (Van Ommen has found that 2% of the forward switching interval is better at producing sharp bands than 1% or 10% (pers. commun.).

We have found that seven different 'ramps' will allow the separation of DNA molecules from 10 to 2000 Kbp in size (Table 2). We find that 1% agarose gels run at 20 °C with a 7 volts/cm potential difference between the electrodes are optimum for yeast chromosome separation.

The optimum duration of the gel run depends upon the physical size of the gel but we find that as a general rule at 7 volts/cm mini gels require 8 h, midi gels 16 h and maxi gels 24 h to complete.

We include a pause interval in our runs of 2% of the forward pulse interval and a forward to reverse time ration of 3:1. Occasionally we also divide our runs into a number of identical cycles, i.e., instead of a mini gel run being a single 8-h cycle, it is composed of two identical 4-h cycles. The switch time interval alters exponentially as the cycle proceeds and generally we allow 40% of the cycle to reach 50% of the forward pulse interval.

Commercially available FIGE systems (DNA STAR and GENE-AID) come complete with software already written to allow separation of a very wide size range of DNA molecules.

FIGE can also be carried out in vertical slab gels (Dawkins et al. 1987) with good results. It has been proposed that this technique presents notable advantages over horizontal slab gels especially with respect to the control of running temperature.

Table 2. Separation of DNA molecules

Sizes of molecules to be separated	Start pulse time (s)	Finish pulse time (s)
10– 100	1	10
10– 200	1	15
10– 300	1	20
10– 500	1	30
20– 800	1	40
20–1000	1	55
20–2000	1	60

REFERENCES

Anand R (1986) Trends Genet 2:278
Brown WRA, Bird AP (1986) Nature 322:477
Burke DT, Carle GF, Olson MV (1987) Science 236:806
Carle GF, Olson MV (1984) Nucl Acids Res 12:5647
Carle GF, Olson MV (1985) Proc Natl Acad Sci USA 82:3756
Carle GF, Frank M, Olson MV (1986) Science 232:65
Chu G, Vollrath D, Davis RW (1986) Science 234:1582
Collins FS, Drumm ML, Cole JL, Lockwood WK, Vande Woude GF, Iannuzzi MC (1987) Science 235:1046
Cook PR (1984) EMBO J 3:1837
Davies KE, Robson KJH (1987) Bio Essays 6:247
Dawkins HJS, Ferrier DJ, Spencer TL (1987) Nucl Acids Res 15:3634
DeGennes PA (1978) J Phys Chem 55:572
De Jonge P, De Jonge FCM, Meijers R, Steensma HY, Scheffers WA (1986) Yeast 2:193
DNA STAR, 1801 University Avenue, Madison, Wisconsin, USA 53705
Estivill X, Farrall M, Scambler PI, Bell GM, Hawley KMF, Lench JN, Bates GP, Kruyer HC, Frederick PA, Stanier P, Watson E, Williamson R, Wainwright B (1987) Nature 326:840
Fangman WL (1978) Nucl Acids Res 5:653
GENE-AID FIGE system, HYBAID. 111–113 Waldegrave Road, Teddington, Middlesex, TW11 8LL, U.K.
Johnston JR, Mortimer RK (1986) Int J Syst Bacteriol 36:569
Kenwrick S, Patterson M, Speer A, Fischbeck K, Davies K (1987) Cell 48:351
Lumpkin OJ, Dejardin P, Zimm BH (1985) Biopolymers 24:1573
Lerman LS, Frisch HL (1982) Biopolymers 21:995
Maniatis T, Fritsch EF, Sambrook J (1982) Molecular cloning. Cold Spring Harbor Press, Cold Spring Harbor, p 65
Olson MV, Loughney K, Hall BD (1979) J Mol Biol 132:387
Patterson M, Kenwrick SJ, Thibodeau S, Faulk K, Mattei MG, Mattei JF, Davies KE (1987) Nucl Acids Res 15:2639
Poustka A, Pohl TM, Barlow DP, Frischauf AM, Lehrach H (1987) Nature 325:353
PULSAPHOR ELECTROPHORESIS Unit 2015-001. LKB-Produktor AB, Box 305, S16126, Broma, Sweden
Schwartz DC, Cantor CR (1984) Cell 37:67
Schwartz DC, Saffran W, Welsh J, Haas R, Goldenberg M, Cantor CR (1983) Cold Spring Harbor Symp Quant Biol 47:189
Serwer P (1981) Anal Biochem 112:351
Snell RG, Wilkins RJ (1986) Nucl Acids Res 14:4401
Van der Ploeg LHT, Schwartz DC, Cantor CR, Borst P (1984a) Cell 37:77
Van der Ploeg LHT, Cornelissen AWCA, Michels PAM, Borst P (1984b) Cell 39:213
Van Ommen GJB, Verkerk JMH (1986) Restriction analysis of chromosomal DNA in a size range up to two million base pairs by pulsed field gradient electrophoresis. In: Davies KE (ed) Human genetic disease: A practical approach. IRL, pp 113–133

2. Isolation of Pure High-Molecular-Weight DNA from the Yeast *Saccharomyces cerevisiae*

The isolation of high molecular weight DNA from yeast depends upon: firstly, protoplasting or spheroplasting the cells and secondly, lysis and the differential separation of the DNA from other cell components. Generally the higher the percentage of cells sphero- or protoplasted, the higher the subsequent efficiency of DNA recovery. However, cells do not need to be fully protoplasted to give quite reasonable yields of DNA. The method which follows is based upon three published protocols and works well with all our laboratory strains of *Saccharomyces cerevisiae*.

a) Protoplasting

Required

1. Protoplast Incubation Medium (PIM) (Allmark et al. 1978)
   0.9 M KCl, 0.015 M $MgSO_4$, 0.0125 M Tris-HCl buffer, pH 7.5 for the pretreatment of cells, β-Mercaptoethanol is added to a final concentration of 1% v/v. During digestion of the cell wall, β-mercaptoethanol was added at 0.1%. In both cases, it should be added just befure use.
2. Protoplast Washing Medium (PWM) (Christensen 1979)
   0.01 M $CaCl_2$, 1 M sorbitol, 0.01 M Tris-HCl, pH 7.5.
   Sterile distilled water.
   Protoplasting enzyme.

Procedure

1. Grow cells overnight in 1 l of YEPD broth at 30 °C with shaking.
2. Harvest by centrifuging at 2000 g for 5 min at 4°, and wash the cell pellets twice in sterile distilled water. Take a small sample and count in a haemocytometer.
3. Resuspend cells to a density of between $10^8 - >10^9$ cells/ml in PIM + 1% v/v β-mercaptoethanol and incubate at 30 °C for 45 min (note 1).
4. Spin down cells in a refrigerated bench top centrifuge at 2000 g for 5 min at 4 °C.
5. Resuspend the cell pellet to the same cell density in PIM + 0.1% v/v β-mercaptoethanol and β-glucuronidase to a final concentration of 3.5% (note 2). Incubate with gentle shaking to prevent the cells sedimenting, at 30 °C for 45 to 90 min (note 3).
6. Gently centrifuge the protoplasts at 1000–2000 g at 4 °C for 10 min in a refrigerated bench-top centrifuge and wash twice in PWM (note 4).

7. Take sample and count. Protoplasts can be stored in PWM for periods up to 2 h at 30 °C and at 4 °C for longer periods (note 5).

b) Protoplast Lysis and DNA Recovery (a modification of the method from Marmur 1961)

Required

1. Lysing Medium: (PLM)
   0.2 M NaCl, 0.1 M EDTA, 2% SDS, 0.05 M Tris-HCl buffer, pH 8.5.
2. Proteinase K (20 mg/ml in sterile distilled water, stored at -20 °C in aliquots, and left at room temperature for 30 min prior to use, to allow thawing and self-digestion.
3. RNase A (20 mg/ml in 0.15 M NaCl. Heated to 80 °C for 10 min to inactivate DNase activity).
4. Phenol, equilibrated to pH 7.5 by saturation with an aqueous solution of 0.10 M Tris-HCl, pH 7.5, 0.001 M EDTA.
5. Chloroform/n-amyl alcohol (24:1).
6. 0.01 M Tris-HCl buffer, pH 7.3, 0.001 M EDTA.

Procedure

1. Protoplasts are suspended to a density of 10^9-10^{10}/ml in PLM. Proteinase K is added to a final concentration of 150 μg/ml and the mixture is incubated at 60 °C for 30-40 min whilst gently shaking (note 6).
2. Add an equal volume of phenol and gently mix the tube contents until a homogeneous white emulsion is formed (note 7). Allow to stand for 5 min at room temperature.
3. Separate phases by centrifuging at 10,000 g for 10 min at 4 °C (note 8).
4. Remove upper aqueous layer and repeat steps 3 and 4.
5. Add an equal volume of chloroform/n-amyl alcohol and mix gently (note 7) and allow to stand for 5 min at room temperature.
6. Separate phases by centrifuging at 10,000 g for 10 min at 4 °C.
7. Remove upper layer and repeat steps 4 and 5.
8. To aqueous layer either:
   A: i) Add $2^1/_2$ volumes of cold absolute alcohol (-20 °C) and mix gently.
   ii) Leave at -20 °C overnight or 30 min at -70 °C.
   iii) Spin down DNA precipitate in a refrigerated bench-top centrifuge at maximum g and 4 °C.
   iv) Pipette off supernatant and wash pellet in 70% alcohol twice.
   v) Add 5-10 ml of 0.001 M EDTA pH 8.3 and leave to solvate.
   Or B: i) Gently layer above aqueous layer, 2 volumes of cold absolute alcohol (-20 °C).

ii) Leave on ice for 15 min and using a glass rod spool off the DNA at interface.
iii) Resolvate spooled DNA in 5-10 ml of 0.001 M EDTA, pH 8.3.
9. To DNA solution add RNase to a final concentration of 0.1 mg/ml and incubate mixture at 37 °C for 1 h.
10. Repeat stages 4, 5 and 6 and 7 and store DNA in aliquots in 0.10 M Tris-HCl buffer, pH 8.3/0.001 M EDTA at 4 °C or -20 °C until required.
11. The concentration and purity of the sample can be estimated by comparing O.D.'s at 230, 260 and 280 nm. (We normally store DNA at a concentration of 1 mg/ml.)

Notes

1. We normally use sterile disposable 50-ml Falcon tubes (Scientific Supplies Ltd, London, England; product no. C3490, or Alpha Laboratories Ltd., Hampshire, England; product no. LW2095), for holding the yeast cells during protoplasting. Alternatively, we used sterilized siliconized glassware.
2. For our standard laboratory strains of *Saccharomyces cerevisiae*, we find that 3.5% w/v of β-glucuronidase is sufficient for protoplasting. However, strain variation in cell wall composition may mean that a higher concentration may have to be used in certain instances. Alternatively, a lower concentration may be sufficient. Note should be taken that β-glucuronidase may not be the best protoplasting enzyme for all strains (see previous section).
3. The time required to protoplast different yeast strains is variable and protoplasting may be followed by placing a drop of the yeast suspension on a slide on the stage of a phase contrast microscope, covering the drop with a coverslip and applying a drop of distilled water to the edge of the coverslip. Protoplasts should be seen to swell and burst on application of the distilled water. Alternatively samples of the protoplasting suspension can be diluted in a cuvette of a spectrophotometer. When the transmission reading after dilution reaches a maximum constant value, when compared to a control of non-protoplasted cells, protoplasting is complete.
4. It is important to wash the protoplasts thoroughly so that no trace of protoplasting enzyme remains. This is particularly important with commercially available β-glucuronidase preparations as considerable contamination with DNases is likely, which can seriously affect the recovery efficiency of the DNA from lysed protoplasts and protoplast transformation efficiency (see later).
Do not pellet the protoplasts too firmly, otherwise resuspension can be difficult and cell lysis can occur. Gentle agitation is advised for protoplast resuspension.

5. For long periods of storage of protoplasts (overnight), we add 0.15 M glucose to the PWM.
6. We find that rather than directly resuspending the protoplast pellet in PLM it is better to resuspend in a small volume of PWM and then add the desired volume of PLM. The main reason for this is that we find that directly adding PLM causes the protoplasts to lyse as a "lump", and DNA recovery is subsequently impaired. If a thick suspension is first produced in PWM, followed by addition of PLM, then lysis is much more homogeneous and DNA recovery is enhanced.

   On addition of PLM the suspension becomes sticky and clearer. This clearing continues gradually during incubation with Proteinase K. Care should be taken not to agitate the tube too much, to prevent shearing of the DNA.
7. The white precipitate is protein and cells membrane complexes. We leave the tubes on a roller for between 5 and 10 min.
8. The protein collects as an interface between the aqueous DNA and phenol layers. Sometimes when there is a heavy protein precipitate it can be centrifuged right through the phenol layer and forms a pellet at the bottom of the tube.

3. Transformation of Yeast: *Saccharomyces cerevisiae*

Transformation of yeast with foreign DNA can be achieved by the use of two different techniques. The first depends on the complete (or partial) protoplasting of yeast cells to allow uptake of the foreign DNA (Hinnen et al. 1978). The second uses intact cells treated with lithium acetate to promote DNA uptake (Ito et al. 1983). The method that yields the best frequency of transformation is strain dependent and also depends upon the nature of the transforming DNA involved. Wide variations among strains in their efficiency of transformation have been observed; e.g. Johnston et al. (1981).

Generally, selection in the yeast recipient strain is by complementation of an auxotrophy, usually produced as a result of deletion of part of a structural gene so that it is not spontaneously reverting, and this complementation can be by foreign material inserted into the plasmid during the experiment or by engineered plasmid-born yeast structural genes. The most common of the latter type are LEU2, HIS3, URA3 and TRP1. A recent review by Parent et al. (1985) lists most types of vectors available for yeast cloning work and a number of other publications are available outlining cloning strategy and vectors used with yeast (Ilger et al. 1979; Botstein and Davis 1982; Hohn and Hinnen 1980; Hinnen and Meyhalk 1982; Broach et al. 1983; Brown and Szostak 1983). It is not the intention here to duplicate these works, but to set out two simple methods which may be used successfully to transform yeast cells.

a) Protoplast (Spheroplast) Transformation (modified from Hinnen et al. 1978)

Required

1. Yeast protoplasts.
2. 1 M Sorbitol/0.01 M Tris HCl buffer, pH 7.5/0.01 M $CaCl_2$ (SCT).
3. 40% w/v Polyethylene glycol 4000/0.01 M $CaCl_2$/0.01 M Tris-HCl, pH 7.5 (PEG).
4. 1 M sorbitol/2% w/v dextrose/2% w/v agar/0.67% w/v yeast nitrogen base (w/o amino acids) (Regeneration agar). Amino acid and organic supplements can be added as required for the strains used and the selection desired.
5. TOP agar, identical to the regeneration agar, but molten and held at 45 °C.
6. SOS broth: 10 ml 2 M sorbitol/6.8 ml of YEPD/0.13 ml of $CaCl_2$/3 ml of H_2O.

Procedure (Assuming an original overnight culture volume of 100 ml)

1. Well-washed protoplasts are suspended in 0.5 ml of SCT and separated into 0.1 ml aliquots in sterile 1.5 ml Eppendorf tubes.
2. DNA solvated in 1 M sorbitol solution is added to all but one of the tubes at the concentration required (0.5–10 μg) (note 1). The final tube serves as a control and receives an equal amount of 1 M sorbitol alone. Tap the base of the tube to ensure thorough mixing.
3. Incubate at room temperature for 15 min, and then add 1 ml of PEG solution, mix, and continue to incubate at room temperature for a further 15 min. Occasionally tap the bottom of the tube to stop the protoplasts sedimenting (note 2).
4. Pellet the protoplasts by centrifugation at 1000–2000 g at room temperature for 5–10 min, and resuspend in 150 μl of SOS broth. Incubate the suspension at 30 °C for 20 min without allowing the cells to sediment.
5. Take 0.1 ml (note 3) of the transformation mixture and add it to 5–10 ml of TOP agar. Mix quickly and immediately pour over the surface of prewarmed (40 °C) regeneration agar plates (1 cm thick).
6. Incubate at 30 °C for 3–5 days and score transformants.

Notes

1. The amount of DNA required to accomplish 'reasonable or measurable' transformation frequency can be variable.
2. At this point, protoplasts can be held for several hours at 4 °C prior to plating without harming the transformation efficiency. At the end of step 3 it is possible to cold-shock the protoplasts to improve the efficiency of DNA uptake. Usually this is done at 0 °C for 30–60 min. The effect of this treatment is variable.
3. Transformation frequencies should be calculated as frequency per regenerated protoplast. This can be determined from the control plates, produced by plating out aliquots of the tube to which no DNA was added. Control plates also give a good idea of contamination. It is often useful to make a dilution of the transformation mixture, and plate out 0.1 ml aliquots of these if it is suspected that the transformation frequency might be high.

    Any transformation mixture not plated out immediately can be stored at 4 °C for short periods. Alternatively, sterile glycerol can be added to a final concentration of 30% v/v and the suspension stored frozen at $-20°$ or -70 °C. These mixture can subsequently be thawed and plated when more transformants are required.

b) Intact Cell Transformation (modified from Ito et al. 1983)

Required

1. 0.01 M Tris-HCl buffer, pH 7.5/0.001 M EDTA (TE).
2. 0.1 M Lithium acetate/Tris-HCl. pH 7.5/0.001 M EDTA (LA).
3. 40% polyethylene glycol 4000 in 0.01 M Tris-HCl buffer, pH 7.5/0.001 M EDTA/0.1 M. Lithium acetate (PEG).

Procedure

1. Grow cells overnight in 50 ml of YEPD broth at 30 °C with shaking (note 1).
2. Harvest cells by centrifugation at 2000 g for 5 min at 4 °C.
3. Wash cells twice in TE buffer.
4. Resuspend cells in 2 ml of LA buffer and incubate at 30 °C for 1 h with gentle shaking.
5. Divide into 0.2 ml aliquots in sterile 1.5 ml Eppendorf tubes and add 0.5 to 20 µg of transforming DNA to nine of the tubes. To the tenth add an equal volume of solvent alone as a control.

6. Add 0.7 ml of PEG solution to each tube and mix by gently inverting the tube several times. Incubate at 30 °C for 1 h.
7. Place the tubes in a water bath at 42 °C for 10 min to heat shock the cells.
8. Allow to cool to room temperature and spread 0.1 ml aliquots directly on to selective media (note 2).
9. Incubate at 30 °C for 3-5 days and score transformants.

Notes

1. Cells should be harvested at a density of 10^7 ml. Each strain that is selected as a recipient in transformation should be characterized for the incubation time required in the media used to arrive at such a cell density before attempting transformation.
2. After stage 8, it is possible to wash the cells in TE and resuspend in TE buffer to plate out. It may be advantageous to make dilutions of the transformation mixture as in the previous method.

4. Plasmid Isolation from Yeast

The method by which plasmids can be isolated from yeast or any other cells depends upon: (a) the volume of culture involved, (b) the purity of the DNA which is ultimately required; i.e. the use to which it will eventually be put, (c) how quickly the plasmid is required, and (d) the size of the plasmid.

Most methods used in the isolation of plasmids from yeast cells are adaptations of those used in the isolation of plasmids from bacterial cells, the principal difference being that yeast protoplasts are the starting material. These methods have been well documented in other texts (Dillon et al. 1985; Perbal 1984; Maniatus et al. 1982), and it is not intended to duplicate them here.

However, a method has been developed (Devenish and Newlon 1982) specifically for the isolation of the yeast ring chromosome III, based on the alkaline degradation of linear DNA, which is also applicable to plasmid isolation, and this is described below:

Required (modified from Devenish and Newlon 1982)

1. 2 M Tris-HCl buffer, pH 7.0.
2. 0.1 M EDTA/1.0 M sorbitol/0.1 M 2-mercaptoethanol/0.2 M Tris-HCl, pH 9.1 (pretreatment buffer).
3. 0.1 M sodium acetate/0.06 M EDTA/1.0 M sorbitol, pH 5.8 (SCE buffer).

4. 0.02 M EDTA/1% w/v SDS/0.05 M Tris-HCl. Adjusted at 23 °C to pH 12.45 (Lysis buffer).
5. Phenol saturated with 3% w/v NaCl.
6. 25% w/v sucrose/0.05 M Tris-HCl, pH 8 (ST).

Procedure

1. Culture cells in 50-100 ml of appropriate supplemented minimal medium to a density of approximately 5×10^7 cells/ml.
2. Harvest by centrifugation at $2000\,g$ for 5 min at 4 °C and wash once in sterile distilled water (note 1).
3. Resuspend cells in 20 ml of pretreatment buffer in a sterile siliconized McCartney bottle and incubate at room temperature for 10 min. Agitate the tube from time to time to prevent the cells sedimenting.
4. Harvest the cells by centrifugation at $2000\,g$ for 5 min at 4 °C and wash twice with 10 ml of SCE buffer. Finally resuspend the cells to a density not exceeding 10^8/ml in SCE.
5. Add Zymolyase 60,000 to a final concentration of $50\,\mu g$/ml and incubate at 37 °C for 25 min.
6. Harvest protoplasts by gentle centrifugation ($1000-2000\,g$ for 5-10 min) and resuspend in 0.5 ml of ST buffer. Place a small Teflon magnetic flea in the McCartney bottle and put on a magnetic stirrer. Stir for 90 s at 100 rpm while adding 9.5 ml of lysis buffer dropwise (note 2).
7. Incubate at 37 °C for 25 min whilst monitoring the pH (with universal indicator strips) and adding 2 M Tris-HCl, pH 7.0 to adjust the pH to between 8.5-8.9.
8. Add NaCl crystals to a final concentration of 3% w/v and incubate at room temperature for 30 min.
9. To the mixture add an equal volume of NaCl (3% w/v) saturated phenol and increase the stirring speed to 300 rpm for 10 s. Then stir at 100 rpm for a further 2 min.
10. Centrifuge at maximum speed in a bench-top centrifuge to separate the two phases, and decant the aqueous layer to a sterile Falcon tube.
11. Add $2^1/_2$ volumes of ice cold ethanol (-20 °C) to the aqueous layer, mix, and stand on ice for 30 min (alternatively -20 or -70 °C). Collect the DNA by centrifugation at 4 °C at maximum g in a bench-top centrifuge.
12. Pour off supernatant, wash pellet in 70% ethanol, vacuum dry and resolvate in TE buffer.

Notes

1. Steps 2-4 refer to protoplasting, and it is probable that any other method could be substituted.
2. The pH of the lysis buffer is critical. Too high a pH may denature the plasmid irreversibly, while too low a pH may not denature linear chromosomal DNA.

5. Rapid Isolation of DNA from Yeast

The method of Devenish and Newlon (1982) allows rapid, efficient isolation of yeast plasmid DNA. However, it does not allow the recovery of high-molecular-weight DNA. On occasion it is useful to be able to obtain relatively small quantities of high-molecular-weight DNA rapidly for restriction analysis, Southern blotting, transformation etc. In such instances, it is laborious and time-consuming to follow the procedure given previously (p. 84). The method described below provides an alternative (Holm et al. 1986).

Required

1. 4.5 M GuHCl (guanadinium hydrochloride), 0.1 M EDTA, 0.15 M NaCl, 0.05% w/v sarcosyl solution, pH 8.0.
2. RNase A (20 mg/ml in 0.15 M NaCl, heated to 80 °C for 10 min to inactivate DNases and stored at -20 °C).
3. Proteinase K (20 mg/ml in sterile distilled water, stored at -20 °C and left at room temperature for 30 min prior to use, to allow thawing and self-digestion).
4. 0.1 M Tris-HCl buffer, pH 8.0, 0.01 M EDTA (10 × TE).
5. Ethanol.
6. 3 M sodium acetate.
7. Zymolyase 60,000.
8. Phenol-chloroform-isoamyl alcohol (25:24:1).
9. 1 M sorbitol, 0.1 M sodium acetate, 0.06 M EDTA, pH 7.0 (SCE buffer).

Procedure

1. Grow cells overnight in 5-20 ml of YEPD broth at 30 °C with shaking.
2. Harvest cells by centrifugation at 2000 g for 5 min at 4 °C and wash twice in sterile distilled water.
3. Resuspend cells in 150 µl of SCE buffer and transfer the suspension to a sterile 1.5-ml Eppendorf tube.

4. Add 10 μl of a mixture of 3 mg/ml Zymolyase 60,000 and 10% v/v 2-mercaptoethanol in SCE buffer and incubate at 37 °C until protoplasting is complete.
5. Spin down protoplasts in microfuge for just long enough to pellet (approximately 2 s).
6. Withdraw supernatant and gently resuspend pellet in 150 μl of GuHCl solution. Incubate at 65 °C for 10 min with occasional agitation.
7. Allow to cool to room temperature and add an equal volume of ice-cold ethanol (−20 °C), and leave for 10 min on ice.
8. Centrifuge at maximum g for 5 min at 4 °C.
9. Resuspend pellet in 0.3 ml of 10 × TE buffer. Stirring may be necessary to facilitate solvation of the pellet.
10. Add 1 μl of RNase solution and incubate the mixture at 37 °C for 1 h.
11. Add 3 μl of proteinase K solution and incubate at 65 °C for 30 min.
12. Extract protein twice with equal volumes of phenol-chloroform-isoamyl alcohol mixture.
13. Add 30 μl of 3 M sodium acetate solution (to adjust final concentration to 0.3 M) and 1 ml of ice-cold ethanol (−20 °C), and leave on ice for 2 h or at −70 °C for 30 min.
14. Pellet DNA by centrifuging at maximum g in a microfuge for 5 min at 4 °C.
15. Wash pellet in cold 70% alcohol and resuspend in the desired buffer.

Recovery: from 5×10^8 haploid cells, the reported recovery is 6 μg of DNA, an estimated efficiency of 75%.

6. RNA Isolation

Several methods exist for the isolation and purification of RNA from yeast cells. Some rely on breaking the cells open by agitation with ballotini beads whilst others rely on simple lysis in phenol. The method to be chosen depends upon the strain used. Generally the efficiency with which a particular yeast strain can be lysed depends upon the strength of the cell wall and it may be necessary to protoplast the cells partially before commencing RNA isolation. Also if RNA degradation proves to be a problem a strain lacking the major secreted RNase of yeast (RNase 3) should be employed. Great care must be taken to ensure that all apparatus and chemicals are sterile and free from RNase contamination. Glassware should preferably be airdried and baked at above 150 °C for at least 6 h. Sterile disposable plastic ware can, however, be treated as RNase-free. It may prove useful to set aside apparatus that is used only for RNA preparation.

Two methods are outlined below, one including agitation with glass beads, the other without. Most methods are similar in nature and can be scaled up or down as required.

a) With Glass Beads (Elion and Warner 1984)

Required

1. 0.01 M Tris-HCl (pH 7.4)/0.1 M EDTA/0.1 M LiCl/1.0% w/v SDS (Soln I).
2. 0.01 M Tris-HCl (pH 7.4)/0.01 M EDTA/0.1 M LiCl/0.2% w/v SDS (Soln II).
3. Phenol equilibrated with TE.
4. 0.4 m diameter sterile acid-washed ballotini beads.
5. 5 M LiCl.
6. 95% ethanol.
7. 70% ethanol/0.05% v/v DECP (diethylpyrocarbonate).

Procedure

1. Grow cells overnight in 10–20 ml of YEPD medium at 30 °C with shaking (note 1).
2. Pour cell suspension into chilled centrifuge tube containing crushed ice (note 2).
3. Harvest cells by centrifugation at 2000 g for 5 min at 4 °C.
4. Wash cell pellet once with 10 ml of ice cold sterile distilled water (note 3).
5. Resuspend pellet in 0.5 ml of cold Soln I and transfer to a sterile Eppendorf tube.
6. Add an equal volume of glass beads and vortex vigorously five times for 30 seconds each time, ensuring that the sample is kept chilled (note 4).
7. Add 1 ml of Soln II and decant supernatant, separate into two equal, approximately 0.75 ml, fractions in clean sterile Eppendorf tubes.
8. Add an equal volume of hot 65 °C TE equilibrated with phenol and vortex vigorously over a period of 10–15 min, ensuring that the temperature remains as close to 65 °C as possible.
9. Separate phases by centrifuging at maximum speed in a microfuge.
10. Remove aqueous layer to a clean sterile Eppendorf tube.
11. Repeat stages 8 and 9 two further times.
12. Precipitate the RNA by addition of 1/25th volume of 5 M LiCl and 2.5 volumes of 95% ethanol and store at −70 °C until required.
13. For analysis, remove an aliquot, wash the RNA in 70% ethanol containing 0.05% diethylpyrocarbonate and resuspend in sterile distilled water.

N.B. The protocol should yield approximately 50 µg of RNA from 20 ml of yeast culture.

Notes

1. The culture should be harvested at mid-logarithmic phase (O.D. 0.6-0.8). To increase the efficiency of RNA isolation it maybe useful to determine the duration of incubation required in the strain being used to reach this point before commencing the procedure for isolation of RNA.
2. Centrifuge tubes should be siliconized if glass, otherwise plastic disposable Falcon tubes can be used.
3. To increase the speed of isolation, stages 2, 3 and 4 can be omitted.
4. At this stage the suspension may be frozen in an acetone-dry ice mixture and stored at $-80\,°C$ for up to a week.

b) Without Glass Beads (Domdey et al. 1984)

(This method extracts only small RNAs (tRNA and 5 srRNA) as other larger RNAs do not penetrate the cell wall matrix.)

Required

1. 0.05 M sodium acetate (pH 5.3)/0.001 M EDTA (Soln I).
2. 10% w/v SDS.
3. Phenol equilibrated with 0.05 M sodium acetate (pH 5.3)/0.001 M EDTA.
4. 5 M sodium acetate (pH 5.3).
5. Phenol-chloroform (1:1).
6. Ethanol.

Method

1. Grow cells overnight in 200-250 ml of YED medium at 30 °C with shaking.
2. Harvest cells by centrifugation at 2000 g for 5 min at 4 °C and wash the cell pellets twice in ice-cold sterile distilled water.
3. Resuspend the cells in 10 ml of ice-cold Soln I in either a siliconized centrifuge tube or a plastic Falcon tube.
4. Vortex the cell suspension vigorously for 3 min in the cold.
5. Add an equal volume of hot (65 °C) phenol and mix thoroughly for 4 min.
6. Rapidly chill the mixture on ice until phenol crystals appear.
7. Centrifuge at 4000 g to separate the phases and decant the aqueous phase to a clean, sterile centrifuge tube.
8. Repeat stages 4, 5, 6 and 7.

9. Extract the aqueous phase with a half volume of phenol/chloroform for 5 min at room temperature and transfer the aqueous layer to a clean sterile centrifuge tube.
10. Bring the aqueous phase to 0.3 M final concentration of sodium acetate by addition of 5 M sodium acetate solution and add 2.5 volumes of absolute ethanol to precipitate the RNA.
11. Store as described for the previous protocol.

REFERENCES

Allmark BM, Morgan AJ, Whittaker PA (1978) Mol Gen Genet 159:297
Botstein D, Davis RW (1982) Principles and practice of recombinant DNA research with yeast. In: Strathern JN, Jones EW, Broach JR (eds) The molecular biology of the yeast *Saccharomyces* II. Metabolism and gene expression. Cold Spring Harbor, New York, pp 607–636
Broach JR, Li YY, Wu LC, Jayaran M (1983) Vectors for high-level inducible expression of cloned genes in yeast. In: Inouye M (ed) Experimental manipulation of gene expression. Academic Press, New York, pp 83–117
Brown PA, Szostak JW (1983) Meth Enzymol 101:278
Christensen BE (1979) Carlsberg Res Commun 44:225
Devenish RJ, Newlon CS (1982) Gene 18:277
Dillon J-AR, Nasim A, Nestman ER (1985) Recombinant DNA methodology. John Wiley, New York
Domdey H, Apostol B, Lin RJ, Newman A, Brody E, Abelson J (1984) Cell 39:611
Elion EA, Warner JR (1984) Cell 39:663
Hohn B, Hinnen AH (1980) Cloning with cosmids in yeasts and E. coli. In: Setlow JK, Hollander A (eds) Genetic engineering II. Academic Press, New York, pp 169–183
Holm C, Meeks-Wagner DW, Fangman WL, Botstein D (1986) Gene 42:169
Hinnen AH, Meyhack B (1982) Curr Top Microbiol Immunol 96:101
Hinnen AH, Hicks JB, Fink GR (1978) Proc Natl Acad Sci USA 75:1929
Ilger C, Farabaugh PJ, Hinnen A, Walsh JM, Fink GR (1979) Transformation of yeast. In: Setlow JK, Hollander A (eds) Genetic engineering: principles and methods I. Plenum, New York, pp 117–132
Ito H, Fukuda Y, Murata K, Kimura A (1983) J Bacteriol 153:163
Johnston J, Hilger F, Mortimer R (1981) Gene 16:325
Maniatis T, Fritsch EF, Sambrook J (1982) Molecular cloning: a laboratory manual. Cold Spring Harbor Laboratory, Cold Spring Harbor, New York
Marmur J (1961) J Mol Biol 3:208
Parent SA, Fenimore CM, Bostian K (1985) Yeast 1:83
Perbal B (1984) A practical guide to molecular cloning. John Wiley, New York

Subject and Author Index

N-Acetylglucosamine 44
N-acetylphenylalanine naphthyl
    ester (APE) 49
Acridine derivatives 35, 36
    acriflavin 37
Adenine 33, 50, 52
    Auxothrophy, ade1, ade2 33, 34
ADP 49
Adriamycin 38
Agarose gels 71–83
Agents producing transversions and
    transitions 35
    Ajam, N. 15
    Mn^{++} 35, 37
    4-nitroquinoline-1-oxide 35
Alcohol dehydrogenase mutants 53
Alkaline phosphatase 57
Alkanes 3
Alkylating agents 35, 58
    EES, EMS, MMS, nitrosoguanidine,
    ICR-170 35, 36, 44, 45, 47–50, 52,
    53, 55, 56, 62
Allmark, BM et al. 85, 97
Allyl alcohol 53, 54
Amicetin 47
α-aminolevulinic acid 51
Aminopterin 52
Amphotericin B 46
Analogues 57
    adenine 58
    4-aminopyrazolopyrimidine 58
    basic amino acids 58
    L-canavanine 58
    L-ethionine 58
    L-methionine 58
Anand, R 74, 80, 81, 84
Aneuploid (industrial yeast strains)
    30
Antibiotics 38
    chloramphenicol 38
    daunomycin 38

erythromycin 38
oligomycin 38
trimethoprim 38, 48
Antisera 44
Asci 13
Asparagine 39
Aspartic acid auxotrophs 55
Auxotrophic tester strains 30

Bacteriophage DNA, as size marker
    78
Ballotini beads 94, 95
Ballou, CE 43
Ballou, L 45
Baranowska, H 39
Base analogues 35
    2-aminopurine 35
    5-bromouracil (5-BU) 35
    5-bromo-deoxyuridine 35
Battaner, E 45
Biotin-deficient mutants 55
Biphasic systems 13
    mineral oil-water 13, 14
    polyethylene glycol-dextran 13
Bird, AP 74, 84
Borst, P 84
Botstein, D 88, 97
Bovine serum albumin 79
Broach, JR 58, 64
Brown, PA 88, 97
Brown, WRA 74, 84
Bruggier, J 35
Burke, DT 74

$CaCl_2$ 46, 65, 66, 85, 89
Cain, KT 42
Canavanine 42
Candida albicans 33
Candida mogii 69
Cantor, CR 75, 84
Carboxymethyl cellulose 68

Carle, GF 71, 72, 73, 82, 84
Casamino acids 52
Cell wall mutants 43
　　easily digested 46
　　fragile mutants 45, 46
　　mannans 43
　　mnn mutants 44
Centromeres 33
　　centromere linkage (determination) 59
Chaput, M 35
Chemical mutagens 3, 35–37
Chloroform/n-amyl alcohol, in DNA recovery 86
Chloromiprimine 38
Christensen, BE 85, 97
Chromosome loss, agents inducing 61
　　acriflavine 61
　　$CdCl_2$ 61
　　$CoCl_2$ 61
　　p-fluorphenylalanine 61
　　methylbenzimidazol-2-yl carbamate (MBC) 61
Chromosome transfer 62
Chromosomes 3, 80, 83
　　human jumping libraries 74
　　rearrangements 74
　　ring chromosome III 91
Chu, G 71, 84
Ciriacy, M 54
Citrate-phosphate buffer 43
Cloning, gene 62
Cohen, JD 58
Cohen, RE 45
Conde, J 42, 62, 63
Contour-clamped homogeneous electric field electrophoresis (CHEF) 71, 72
Cooper, TG 57, 58
Cornellisson, AWCA 84
Culbertson, MR 49, 51
Cycloheximide 42
Cytoduction 42

Davies, KE 74, 84
Davis, RW 62, 63, 83, 88, 97
Dawes, IW 12, 15
Dawkins, HJS et al. 83, 85
De Jonge, P (and De Jonge, FCM) 73, 84
Deaminating agents 35
　　hydroxylamine 35
　　nitrous acid (NA) 35, 36

Debaryomyces sp 10
DeGennes, PA 71, 80, 84
Dejardin, P 84
Density gradients 13
　　renografin 13, 15
　　urografin 13, 14, 15
Devenish, RJ 91, 93, 97
4,6-Diamino 2-phenylindole (DAPI) 37
Diethylpyrocarbonate (DECP) 95
Dillon, J-AR 91, 97
Dimethylsulfoxide 65, 66
Direct manipulation of DNA and RNA 71–97
Dissecting needles 17, 18, 22
Dissection chamber 16
Dithiothreitol 65, 80
DNA Star system 72, 83
DNAases (role in degrading high-molecular weight DNA) 76, 77
Domdey, H 96, 97
Drop-overlay 4
Drying, as mutagen 33
Duchenne muscular dystrophy 74
Dutcher, SK 62

EDTA 75–78, 86, 87, 90–96
Elion, EA 95, 97
Emetine 47, 48
Estiville, X et al. 74, 84
Ethanol 7, 44, 55, 56, 92–96
Ether, diethyl 11
　　killing of vegetative cells 11
Ethidium bromide 35–37, 82
Evans, IH 57, 58

Fangman, WL 71, 79, 84
Fast Garnet GBC (a diazonium salt) 49
Fatty acid-requiring mutants 49
Field, C 58
Field-inversion gel electrophoresis (FIGE) 71–74, 82, 83
Fink, GR 42, 62, 63
Fischbeck, K 84
Fluorescein 45
　　fluorescein-labelled wheat germ agglutinin (fl-WGA) 45
Fluorescence-activated cell sorting 44
p-fluorphenylalanine 41
Fogel, S 25, 62, 63
Fowell, RR 8

Subject and Author Index

Fragile mutants 45, 46
Fragile X syndrome 74
Frank, M 84
Freezing mixtures (dry iceacetone) 34
Frisch, HL 71, 80, 84
Fructose-1,6-bisphosphatase 56
Fukuda, H 69
Fusidic acid 47

Gancedo, C 57
Gelatin 12–14, 69, 80
Gels, loading 81
  running 81
Gene-Aid FIGE system 72, 83
Genes, auxotrophic 59
  trp1, pet18, met14 59
Gjermansen, C 62
Glucanases 76
Glucose 9, 14, 18, 20–22, 30–32, 34, 37–39, 41–56, 65–69
Glycerol 7, 29, 37–39, 41, 52, 55, 56, 67
Glycolytic cycle mutants 53
  alcohol dehydrogenase 53
  hexokinase 53
  mutants in galactose utilization 53
  PEP-carboxykinase 53
  phosphofructokinase 53
  pyruvate carboxylase 53
Goldenberg, M 84
Goldenthal, MJ 57, 58
Gorin, PAJ 45
Gorodkowa agar 10
Greeneberg, ML 51
Gregory, KF 46, 47
Groves, DP 68
Guanidinium hydrochloride (GuHC1) 93, 94

Haas, R 84
Hadjiolov, AA 45
Hansen, RJ 56, 57
Hansenula capsulata 68
Hansenula polymorpha 3
Hansenula wingei (H.Canadensis) 68
Hardie, ID 12
Hawthorne, DC 59, 60, 62
Haynes, RH 52, 53
Heme 51
Hexokinase 56
Hide powder azure 49
Hinnen, AH 88, 89, 97

Hinze, H 57
Histidine 43
Hohn, B 88, 97
Holm, C 93, 97
Holmberg, S 62
Holzer, H 57
Homothallism 11
Human X chromosome 74
Human monogenic diseases 74
Hypoxanthine 35

Ilger, C 88, 97
Imidazole 56
Inositol-excreting mutants 50
Inositol-requiring mutants 49
Invertase 57
Isolation of pure high-molecular weight DNA from yeast 85–88
  DNA recovery 86
  protoplast incubation medium 85
  protoplast lysis 85, 86
  protoplast washing medium 85
Ito, H 88, 89, 90, 97

Johnston, JR 73, 77, 84
Jones, EW 49, 58, 64

"Kamikaze" mutants (antibiotic-resistant) 47
Kar1 mutation, in mapping 62, 64
Karst, F 51, 52
Kenwrick, S 74, 84
KH_2PO_4 39, 41
Kielland-Brandt, MC 62
Killer factor 57
  ds-RNA 57
  Virus-like particles (VLP) 57
Kimura, A 69
Kluyveromyces lactis 3, 43, 44
Krebs cycle 55

Lactate 7, 37, 55
Lautsten, O 3, 4
Lerman, LS 71, 80, 84
Leucosporidium scottii 3
Lin, Y 35
Linkage 3
Lipase 46
Lithium acetate 65, 88, 90
Lithium chloride (Li Cl) 95
Little, JG 52, 53
Littlewood, BS 48
Lodder, J 10

Lumpkin, OJ 71, 80, 84
Lysine, lysine auxotrophs 39

Magee, PT 35
Malate dehydrogenase 56
Malt extract agar 10, 13, 14
Maltase 57
Maltose 57
Maniatis, T 79, 84
Mapping (gene) 58
   chromosome loss 58
   chromosome transfer 58
   fine-structure mapping 58
   mitotic crossing-over 58
   mitotic gene conversion 58
   trisomic analysis 58
Marmur, J 58, 86, 97
Mating 4
   rare-mating 5
Mating-type switching 7
   determination by cross-stamping 28–30
McClary's medium (sporulation) 8, 9
Mehta, H 46, 47
Meiosis 3
Melham, T 55
Membranes 49
   phospholipids as components 49
Menadione (Vitamin K_3) 35
2-Mercaptoethanol 65, 75, 77, 85, 91, 94
N-Methyl-N'-nitrosoguanidine (MNNG) 36
Meyhalk, B 88, 97
$MgSO_4 \cdot 7H_2O$ 39, 77, 85
Michels, PAM 84
Microforge 23
Microloops 22, 23
Micromanipulation 4, 18–20, 23, 24, 28
   spore-cell pairing 5
   spore-spore pairing 5
Miller, R 69
Mineral oil 12, 13
Mitochondria, mitochondrial DNA 36, 37
Mitotic mapping 61
   mitotic chromosome loss 61
   mitotic crossing-over 61
Mitotic recombination 33, 61
$MnCl_2$ 56
Mortimer, RK 25, 60, 62–64, 73, 77, 84

Mouse genome 74
Mushroom enzyme, preparation 25, 26
Mutagenesis 30–58
Mutants auxotrophic for 2-deoxy-5'-monophosphate (dTMP) 52, 53
Mutants defective in nuclear fusion (kar) 42, 64
   Strain JC1 (his4 ade2 can1 nys^R p^-) 42
   Strain GF4836 (leu1 thr1 p^+) 43
Mutations increasing chromosome loss (cdc6, cdc14, chl1, rad52, spoll) 61
Mutator mutants 39–42 -
   Strain carrying trp5-48 his5-2 arg 4-17 lys1 ade2-1 40

NaCl 44, 86, 92, 93
NADH 56
$NaHCO_3$ 56
α-Naphthol 49
Newlon, CS 91, 93, 97
Nilsson-Tillgren, T 62
Novick, P 58
Novozyme 75, 76
Nystatin 43, 51

Oliver, SG 68, 69
Olson, MV 72, 75, 82, 84
Orange-G (dye) 79
Orthogonal-field alternation gel electrophoresis (OFAGE) 71–74, 82
Osmotic-remedial mutants 42
   KC1 42
Osmotolerant yeasts 69
OSY-glucose medium 66
Oxidative phosphorylation 55
Oxygen, effects on survival 33

PAGE-SDS electrophoresis 54
Palleroni, N 8
Parent, SA 73
Park, FJ 49
Parks, LW 51, 52
Partridge, RM 35
Patterson, M et al., 74, 84
PEP4 mutants 48
   alkaline phosphatase 48
   carboxypeptidase Y 48
   proteinase A 48
   RNAase 48

Subject and Author Index

PEP4 mutants
 vacuolar hydrolases 48
Peptidase 46
Perbal, B 91, 97
Perea, J 57
Petersen, JGL 62
Petite mutants, petite mutation 7, 36, 37, 57, 68
 failure to metabolize sugars 57
 using acriflavin 7
 using ethidium bromide 7
Phelps, J 54
Phenol 86–88, 93–97
Phenol-chloroform 96, 97
Phenol-chloroform-isoamyl alcohol 93, 94
Phosphatase (genes) 53
 PHO80 gene 53
Phosphoenolpyruvate 56
Phosphogluconate dehydrogenase 53
 mutants
Pichia sp, *Pichia pinus* 3, 10
Piedra, D 67, 69
Plasmids, plasmid isolation from yeast 3, 91–93
Plate holders 23, 24
 Goldsmiths' College 24
 Lawrence Engineering 23
PMSF (phenylmethylsulfonyl fluoride) 79
Polyethylene glycol (PEG) 45, 65–69, 89–91
Polyvinyl alcohol 68
Post-meiotic segrgation 22–25
 Queen Mary College method 22
Potassium 55
 acetate 55
 KC1 as stabilizer 8, 77, 85
 PO_4 in buffers 8, 77
Potter-Elvehjem homogenizer 12, 13
Poustlea, A 74, 84
Powdered glass 13
Prazmo, W 39
Preparation of yeast DNA molecules 75–77
Proteases 3, 76
Proteinase K 75, 76, 80, 86, 88, 93, 94
Protoplasts 8, 12, 76, 77, 85–90, 91, 92, 93, 94
 formation and fusion 64–69
 regeneration, in solid medium 64
Pulsaphor system 72, 74
Pulsed-field gel electrophoresis 71–84

Putrament, A 39
Pyruvate carboxylase mutants (pyc) 55

Rabbits, for raising of antisera 44
Raffinose 10
Random spore isolation 10
Recombinant DNA 3
Reiner, B 51
Replica plating 26–28, 29
 using filter paper 27
 using velvets 26
Reptation 71, 72
Restriction enzymes 3, 74, 80
 Not1, Sfi1 74
Rhoads, DH 35
Rhodosporidium toruloides 3
Rhodotorula sp 3
Rifampicin 46
RNA isolation (with and without glass beads) 94–97
RNAase, in recovery of DNA 86, 87, 93, 94
Robson, KJH 74, 84

Saccharomyces cerevisiae 3, 7, 44, 48, 49, 57, 64, 67–69, 79, 82, 85, 87, 88
Saccharomyces diastaticus 10
Saccharomyces montanus 68
Sachcharomyces rosel (Torulaspora delbrueckii) 68
Saccharomycodes ludwigii 3
Saccharomycopsis capsularis 10
Saccharomycopsis fibuligera 10
Saffran, W 84
Saracheck, A 33, 35
Sarcosyl 93
Schambart, DHJ 57
Schekman, R 58
Schild, D 64
Schizosaccharomyces pombe 3
Schlessinger, D 45
Schopfer's medium, for starvation 37
Schwartz, DC 71, 75, 84
Schwartzhof, RH 35
Second-division segregation 59
Secretory mutants 57
 Class A 57
 Class B 57
Sequencing, DNA 62
Serwer, P 71, 84
Sherman, F 16
Shuttle vectors (yeast/*E.coli*) 74

Size standards for determination of large DNA molecules (oligomeric ladders), bacteriophages T2, T7, G 78, 79
  yeast chromosomal DNA (from strain X2180) 79
Skim milk 34
Smith, ARW 25, 62, 63
Snail enzyme (glusulase, helicase, β-glucuronidase 12, 14, 18, 46, 65, 75, 76, 87
Snell, RG 82, 84
Sodium acetate 91, 93, 94, 96
Sodium citrate 36
Sodium glutamate 34
Sodium lauryl sarcosinate 75
Sodium pyruvate 55
Sodium thiosulfate ($Na_2S_2O_3$) 36, 45
Sorbitol 45, 68, 85, 91
Speer, A 84
Spencer, DM 8, 25, 62, 63, 68, 69
Spencer, JFT 8, 25, 62, 63, 68, 69
Spermidine 80
Sporulation 5, 12, 15
Starvation 37
Steinberg, CM 42
Sterol-requiring mutants 51
  cholesterol 51
  ergosterol 51
Strathern, JN 58, 64
Struhl, K 62, 63
Styles, C 62, 63
Sucrose 45, 92
Sulfanilamide 52
Super-triploid method (mapping) 60
Svoboda, A 67, 69
Szostak, JW 88, 97

Taylor, FR 51, 52
Temperature-sensitive mutants 42, 51
  cold-sensitive mutants 42
Tetrad analysis 16–25, 58–60
  spore-cell pairing 21
  spore-spore pairing 21
Thioglycollic acid 65
Tongue depressors, for cross-stamping 28
Transformation 88–91
  of intact cells 90, 91
  of protoplasts 89–90
Transport mutants 57
  amino acid analogs 57
  heavy metals 57
  and resistance to inhibitors 57
Tris, Tris-HC1 buffer 46, 47, 65, 75, 78, 85–87, 89, 90–95
Tris-acetate buffer 77
Tris-borate buffer 77
Trisomic analysis, for assignment of genes to chromosomes 60
Trypanosoma brucei 74
Trypsin 46

Ultraviolet irradiation 3, 30, 32, 33, 46, 47, 51, 58, 62
Uracil 39

Van de Poll 56, 57
Van der Ploeg, LHT 74, 84
Van Ommen, GJB 74, 78, 80, 84
Venkov, P 45, 46
Verkerk, JMH 74, 78, 80, 84
Vertical slab gels, for FIGE separation technique 83
Vezinhet, F 15
Vollrath, D 84
Von Borstel, RC 39, 42

Welsh, J 84
Wetter, LR 64
Whelan, WL 33, 35
Whittington-Vaughan, P 69
Wickner, RB 60, 62
Wilkie, D 57, 58
Wilkins, RJ 82, 84
Wills, C 54, 55
Winge, O 3, 4
Wright, JF 15

X-ray 3, 33, 58, 62

Yarrowia lipolytica 3
Yeast extract 9, 13, 18, 20, 22, 28, 30–32, 34, 37–39, 41–56, 65–68
Yeast Genetics Stock Center 64
Yeast-nitrogen base 4, 7, 28, 30, 31, 34, 37, 39, 55, 89

Zimm, BH 84
Zubenko, GS 49
Zygosaccharomyces bailii 65
Zygosaccharomyces cidri 69
Zygosaccharomyces rouxii 65
Zygotes 7
Zymolyase 12, 13, 18, 25, 65, 75, 76, 92, 93, 94